Into Dental Practice
A Working Guide

Into Dental Practice
A Working Guide

Published by the British Dental Association
64 Wimpole Street, London W1M 8AL

ISBN 0 904588 28 9

Typeset by E T Heron (Print) Ltd, Essex
Printed in England by Eyre & Spottiswoode Ltd, London and Margate

Contents

Preface

Vocational training in general dental practice came of age in 1988 when realistic funding was agreed between the profession and the Department of Health. Already over half of the output of dental schools are involved in some form of vocational training. The overall plan, whereby recent graduates enter practices which have been approved by universities and trainees attend an educational programme throughout the year, has been tried and tested. Evolutionary experimentation has been encouraged both in the training practices and on the day-release courses.

Although vocational training has as its primary ojective guidance during the metamorphosis of student to confident clinician, trainers have found that there have been major benefits both to themselves and to their practices as a result of their involvement with VT.

This compendium is of interest to trainers, trainees and course organisers. Matters which were unclear have now been defined —the trainee's contract is a good example. The difficulties that new graduates find in understanding the health service and the management of a surgery are clearly addressed and the practicalities of providing clinical care in general practice surroundings are discussed.

This is a companion to the more formal guidance given by the Committee on Vocational Training. All of the authors have been involved in vocational training in one way or another. Vocational training is not about repeating or replacing any part of the undergraduate curriculum. It is about accelerating the production of quality in care for patients in general practice and this is what these chapters will help you as trainers, and you as trainees, to achieve.

Gordon Fordyce
Chairman
Committee on Vocational Training

Introduction

The benefits of the new contract, introduced on October 1, 1990, for patients and dentists have still to be proven. However, the benefits for patients and dentists of vocational training have already been shown beyond reasonable doubt. The success of the national voluntary vocational training scheme is a tribute to course organisers, trainers and trainees alike. There had been guinea pigs before them—the Barnet and Guildford schemes run by Edgar Gordon and John Brookman had been models for a dozen years and more. Nevertheless the introduction of a national scheme on January 1, 1988 was still an act of faith. Here the dream of Dr Desmond Greer Walker became reality under the guidance of Gordon Fordyce, his successor as postgraduate dental dean for the four Thames regions and the chairman of the Committee on Vocational Training (CVT). The CVT was set up as an intermediary between the health departments, the GDSC, deans and regional advisers to lay down the ground rules. Now its role is largely one of monitoring progress. This includes a programme of visits to regional training schemes. A team of three or four visitors spend 2 days talking to the dean, course organisers, trainers and trainees visiting training practices, and sitting in on the day-release course. The visits have higlighted the good and the not-so-good aspects of regional schemes, but suggestions for improvement have been well-received. The CVT's newsletter has been used to promote and share the good ideas. A course directory for prospective trainees has also been drawn up by the CVT, in association with the BDA, and the CVT's secretariat at the British Postgraduate Medical Federation acts as a clearing house, matching trainees with vacant training places.

This clearing house role will become all the more important when vocational training becomes mandatory. It is still expected that a European Community Directive will tell member states to introduce dental vocational training for entry to their social security systems by 1997 at the latest. The UK will get there sooner than most, but there are obstacles on the way. There are not yet enough places for all 600 or so new graduates wishing to enter the GDS, not to mention the graduates from other EC countries coming to this country, and those dentists who wish to transfer from the hospital or community services to general practice.

At whatever date vocational training for general practice does become mandatory, the national scheme will need to be more flexible in its attitudes towards experienced practitioners. The present voluntary scheme is restricted to newly-qualified dentists. Dentists with more than one year's post-qualification experience are generally not admitted to the scheme, although they are occasionally allowed to attend the day release course component. In England and Wales course organisers or postgraduate dental deans usually ask the Committee on Vocational Training to make a decision on an experienced practitioners's eligibility to join a scheme on a dentist by dentist basis. New graduates should not be forced to make irrevocable career decisions on graduation, which would be the effect of coupling mandatory vocational training with a training scheme which did not accept dentists with more than one year's experience. The priority of the organisers of the present scheme is to meet the 1993 target for enough places for new graduates (this is the earliest that VT could be made mandatory: the BDA's target is 1995) but after that date there would need to be some slack in the system to cater for those dentists wishing to change career paths, and who would need a certificate of completion of vocational training, or of equivalent experience, in order to enter NHS general practice. Without a mechanism for establishing equivalent experience, entrants to general dental practice would, apparently, have no option but to go through VT in its entirety.

At present the Committee for Vocational Training (England and Wales), the Scottish Dental Vocational Training Committee, and the Northern Ireland Council for Postgraduate Medical Education accredit vocational training schemes. Wales is seeking increasing autonomy in educational matters, so there soon may be four national bodies within the UK responsible for the management of vocational training schemes.

The CVT is currently drawing up what it believes should be the definitive distribution of training schemes. Every region should have at least one scheme, but it is difficult to lure graduates away from the area of their own dental school. The regions in England without a school—Oxford, Wessex and East Anglia—are particularly aware of this problem.

There will always be some new graduates who will prefer to go straight into associateships. Sadly, as the BDA and the defence societies know all too well, there is no comparison between the experiences of these new graduates and their trainee contemporaries. The CVT's survey of new graduates shows just how lacking is their clinical experience of commonly-encountered problems in general practice. If this is not enough to show the value of VT before independent practice, the selection of training practices and trainers offers further protection to new graduates at the most vulnerable stage of their career. But for those young dentists who still fancy their chances of making money in the first year of practice (often unrealistically) there will be a financial penalty for avoiding VT. Under the new contract, anyone entering a an FHSA/Health Board list after April 1, 1991 will receive a lower level of continuing care payments, unless they have undertaken vocational training first.

Differential payments and the imminence of mandatory VT mean that no undergraduate, and no practice looking to take on a new graduate can afford to ignore vocational training. This compendium of *BDJ* articles is as good a starting point as any for finding out what the scheme is all about. The articles are written by people in the know, including two postgraduate deans, three course organisers and two trainees, but they cover far more than the nuts and bolts of the scheme. Contracts of employment for staff, cross-infection control, practice management tips as well as the clinical articles should be of interest to all GDPs, whether or not they are currently involved in VT.

Mark Paulson
Executive Secretary
British Dental Association

1

CVs and Letters of Application for Posts

P. S. Rothwell

Competition for posts in dentistry is very keen, and often the decision to interview or reject outright is made on the basis of information provided by the letter of application or a curriculum vitae. It is important that these are used to their full effect and represent the candidate in the best possible light. This chapter is intended to guide the final-year dental student or the newly-qualified dentist in the preparation of such documents.

Not too long ago it was common for advertisements for posts as assistants or associates in general dental practice to attract only one or two applications, and, therefore, a principal's task was relatively simple. There was more than adequate time to see each applicant and no need to become involved in shortlisting the potentially suitable. Today, the situation is different.

A principal who advertises through a box number in the *British Dental Journal* may receive more than 20 applications, and often these will arrive in one package. This creates a situation more like that for applications for salaried positions, where an appointments committee studies all the applications and shortlists the best candidates. Shortlisting is assisted by curriculum vitaes, and because of the changed situation it is becoming more common for one to be required for applications for posts in general dental practice.

Why is a CV necessary when all the information could be included in a letter of application? For two main reasons; first, letters of application are individually written for each application whereas a CV requires only one preparation and it can then be photocopied. Secondly, letters need to be written in a prose style whereas CVs are produced in a tabular form. One page of CV can include information that would need several pages of rambling script in a letter. Consequently, the combination of a letter and a CV offers the potential employer a very rapid and easy-to-assess overview of an applicant's suitability. Initially, therefore, the aim of the letter of application and the CV is to give the best chance of being selected for interview. This chapter describes how to prepare these important documents.

Often, dental students or the newly qualified think that their relatively short lives are too uninteresting to justify a CV. In some cases, sadly, this is true. Many have done little except study for and pass exams. On the other hand, others have experienced more in their first 22 years or so than many people do in a lifetime. When helping students to prepare a CV, the problem, on occasions, is not one of searching for sufficient material but one of how to condense experiences and accomplishments within a concise CV.

Whereas the CV is a formal document, the covering letter is a personal communication. However, it is just as important in its own way and requires the same degree of care in its compilation. Any examples of 'dos and don'ts' quoted in these articles are based on my experiences and those of colleagues. If you think 'no one would be stupid enough to do that', you are mistaken—they have (and still do).

The letter of application

Because it is essentially a personal document, a good handwritten letter is more attractive than one which is typed, particularly for GDP posts. The notepaper does not have to be the most expensive on the market nor printed with the family crest, but a perforated sheet torn out from a notebook is just not good enough. It should be neat and free from crossings-out, coffee stains, spelling mistakes or other evidence of careless habits.

As the first thing seen by the advertising principal will be the envelope, make sure that it too is of reasonable quality and that it has a postage stamp on it, rather than the imprint of a dental school's (or anyone else's) franking machine. If you make such an open declaration that you are prepared to 'borrow' someone's franking machine, a principal might just wonder whether you are the sort of person who would be equally prepared to 'borrow' from his or her practice. Details matter and cost little in terms of time or money.

The covering letter should include, briefly:
- why you have applied for that post in that area of the country;
- what features quoted in the advertisement (such as 'family practice', 'hygienist' and 'preventive') particularly attracted you;
- why you prefer general practice rather than other branches of dentistry;
- if appointed, how long you would anticipate staying in the post;
- how the post fits into your career plan. If you have no particular career plan, say so, but explain why you think this post, as such, would be valuable.

Remember, this is a letter of application to be a 'learning guest', because that, in a nutshell, describes a first post in dental practice. So keep the letter short and simple, without outlining your philosophy of the future of dental practices in this country or how you think they should be run. Such comments engender apprehension in a principal rather than admiration.

You are hoping for an interview, but never write in terms of 'I should like to see your practice . . . ' as this can be interpreted (and has been) as 'I should like to view your practice to see if it would be good enough for me to work in . . . '. Word it, for example, 'If you consider my application to be of interest, I am available for interview at any time . . . '. It is worth investing in a first-class stamp and enclosing a self-addressed envelope. It probably will not be used (to save time, a principal will probably 'phone) but it is a thoughtful gesture.

Overall, the letter should give an impact of care and attention. If you cannot spell, buy a dictionary. Nothing creates a worse impression than a letter that looks and reads

Telephone   0111 123 6789 (evening)
             0111 123 4567 (day)

25 Every Street,
Newtown,
Dentshire AB12 YZ34.

18th December, 1990.

The Advertiser,
Box No. 12345,
British Dental Journal,
Professional and Scientific Publications,
BMA House,
Tavistock Square,
London WClH 9JR.

Dear Sir or Madam,

I wish to apply for the post of Associate which you advertise in the 'British Dental Journal' (10th December, 1990).

I qualified recently from Newtown University, and I wish to pursue a career in general dental practice rather than in other branches of dentistry. I am anxious to work in a suburban family practice of the type you describe because I think this would offer the most comprehensive experience. In particular, I enjoy treating children. I should like to work in the area of your practice because, as you will see from the enclosed Curriculum Vitae, I am a keen dinghy sailor and there are several excellent clubs within easy travelling distance.

I have no firm future personal plans or commitments, and if appointed and satisfactory, I would be able to offer a reasonably long period of service. I should be pleased to attend for interview, and am available at any time.

I enclose a prepaid, addressed envelope, but can be contacted by telephone as above.

The following have agreed to supply references if required.

Mr J. Knowitall, MDS, FDS,   Mr A. S. Practice, BDS,
Senior Lecturer in Prosthetics,   P/time General Dental Practitioner,
Department of Prosthetics,   Department of Conservation,

Address for both:
The Dental School,
Oral Way,
Newtown,
Dentshire AB13 PQ34

Thank you for your attention.
Yours faithfully,

(signature)

JENNIFER CARIES (Ms)

as though it has been scribbled over, or partly in, your breakfast. It is an insult to the principal because it implies that you think him or her too unimportant to warrant your best effort. Do not forget that scrappy letters hold the world's speed record for reaching the waste-paper basket, and if you make no effort when writing the application, do not expect the principal to make the effort to reply. You are, or about to become, a professional person applying for a professional post. This is an excellent moment to start getting used to acting like one.

Layout is important. An example of a covering letter is on page 2. (It has not been shown handwritten for ease of publication) and several points should be noted.

● Never put your name above the address and date (top right-hand corner).

● When replying to a box number, address it to 'The Advertiser'.

● If the advertiser is anonymous, open 'Dear Sir or Madam' (many applicants write only 'Dear Sir', and this can offend women principals).

● If you open 'Dear Sir or Madam', you must end 'Yours faithfully'. 'Yours sincerely' is used only after opening by name, such as 'Dear Mr Johnson'.

● Give daytime and evening telephone numbers if you have both.

● State that the referees have agreed to be approached.

● Always print your name under your signature and add Mr, Mrs, Miss or Ms.

The curriculum vitae

Ideally, this should be typed. If necessary, buy professional help. It will cost little enough from a secretarial bureau, or secretaries whom you know may be prepared to do this for you 'after hours', but always offer to pay them. Avoid preparing your own CV on a personal computer unless you have access to a daisywheel or laser printer—dot matrix printers produce dot matrix images. Also, in my experience, domestic wordprocessing, spelling mistakes and typographical errors seem to go hand in hand. Top-quality initial printing produces the best photocopies and you may need many.

CV layouts vary according to the type of post involved, but do not differ substantially. A sample CV is shown on page 4. It is assumed that this is for a first post. Eventually, it would be necessary to add details of present and previous posts, papers published, lectures given and so on, but this will not apply immediately after qualification. In general, aim to put the information on one sheet of A4 paper. Routine type size may cause this length to be exceeded slightly, but with modern photocopiers it is possible to reduce the copy to an A4 sheet. However, the finished reproduction should have print of easily-read size.

'Personal details'

● Full names must be given with the surname underlined or in bold type. To the English-speaking reader, the surname of other nationals may not be obvious, or, of course, vice versa. It is for this reason that gender is also given.

● Date of birth. There is no need to put date of birth and age, which would suggest that the reader cannot perform simple arithmetic.

● Marital status. If you have children state their sexes and ages.

● Address(es). As a student, put both your home and term-time addresses, together with all telephone numbers.

Education

There is no need to start from 'Miss Egbert's Kindergarten'. List only secondary schools and university. The dates are best placed on the right, and in a neat column so that the reader can quickly scan the chronology and detect any unaccounted periods. Your CV must account for every year because omissions may suggest periods of serious illness or other significant factors. Do not leave the reader guessing.

Academic achievements

It is questionable whether you need to list every 'O' and 'A' level. Do so if you wish, but it takes up a lot of space. The numbers obtained, grades and dates alone should suffice, unless, for instance, you have an 'A' in advanced-level Spanish and you are applying for a post in Madrid.

Prizes and distinctions

Include all major prizes and distinctions, whether they were achieved at school (for example, science prize—sixth form), university (for example, Honours Anatomy, second BDS) or as part of an interest (for example, Duke of Edinburgh Gold Award).

Positions of responsibility

Most of these are onerous, unpaid, thankless and indicate that you are not one of the apathetic majority. Include secretary of societies, editorship of student magazines, captain in sports, year representative and the like.

Interests and activities

These should be interests in which you play an active part. Do not just quote a string of topics resembling replies at a beauty contest. If possible, divide them into groups such as sport and music.

Other details

You can add a section such as this if you have any points to make that are not covered elsewhere. Detail of elective periods spent abroad during the undergraduate curriculum is an example, or the section could be used to explain 'non-standard' paths to qualification by someone who has previously obtained a qualification in another discipline. Do not list all your vacation jobs unless they are particularly relevant to the qualities a principal might be seeking. For instance, periods as a 'nanny' would suggest that you have an empathy with children and this would be valuable in a 'family' practice.

References

Whether you name referees in the CV or in the letter of application is a matter of choice. I prefer naming them in the letter because you may wish to use alternative referees when applying for different posts. Referees should be those with knowledge of you most relevant to the post for which you are applying. Thus, for hospital posts, the best referees are from hospital staff, whereas for a general dental practice post, it is possibly more germane to ask a GDP part-timer, together with a senior member of the academic staff, to be referees.

Whoever your referees, before you quote them always extend to them the courtesy of asking permission. Not only is this good manners, but it will also alert them to keep a note of details for future use, because sometimes one is asked for references some years after a student has qualified.

A good CV will give a principal not only an overview of your life and evidence of your potential for the post advertised, but also offer clues to non-professional topics that can be explored at interview. The more a principal knows beforehand, the better he or she is prepared for the interview.

Write out your CV in rough first and then leave it for a couple of days, re-read it and redraft it. After two or three attempts, you will have increased the information and reduced the length. Only when you are satisfied that you have condensed the maximum information into the minimum number of words should you produce the final copy.

When applying for a post, you are in a competitive situation. Your CV should indicate the relevant talents or qualities you have which put you ahead of the field and it should make a principal think 'I must see this applicant'.

CURRICULUM VITAE

PERSONAL DETAILS

Name	Jennifer CARIES	
Sex	Female	
Date of Birth	6/2/68	
Address	Term	Home
	25 Every Street,	25 The Cedars,
	Newtown,	Bigtown
	Dentshire AB12 YZ34.	Mouthshire CD23 XG19.
Telephone	0111 123 4567 (day)	0123 341254
	0111 123 6789 (evening)	
Marital Status	Single	

EDUCATION

Bigtown City School	1979–1984
Bigtown College of Further Education	1984–1986
Newtown University	1986–1990

ACADEMIC ACHIEVEMENTS

9 '0' levels (4-A, 3-B, 2-C)	1984
3 'A' levels (2-B, 1-C)	1986
BDS (Newtown)	1990

PRIZES AND DISTINCTIONS

Physics Prize (Bigtown College)	1986
Duke of Edinburgh Gold Award	1986
Distinction 2nd BDS	1987
'Sharp Scalpel' prize in Oral Surgery—final BDS	1990

POSITIONS OF RESPONSIBILITY

Section Leader—Bigtown College Himalayan expedition	1985
Secretary—Dental Students' Sports Committee	1988–1990
Treasurer—Newtown University Music Society	1987–1989

INTERESTS AND ACTIVITIES

Sport	—	Member Seasoch Sailing Club
	—	Newtown University ladies 1st squash team
Music	—	Piano (grade 8), Violin (grade 6)
	—	Former member Bigtown Choral Society
Photography—		Develop and print black/white prints and colour slides
	—	Member Newtown University Photographic Society

OTHER DETAILS

In 1990 I undertook one month's student elective in Kokoland, where I worked with a Government District Dental Officer to gain experience of dental health problems in a Third World country.

During vacations, I work as assistant in a home for the disabled (Bigtown).

2

Preparing for Interviews

P. S. Rothwell

An interview can range from the formal 'across a desk' type to the very informal, where the applicant spends time in a dental practice during a normal day and talks with the principal while he or she is working. Clearly, it is not possible to offer guidelines for every circumstance, but whatever the method, similar topics are covered. Regardless of the format, preparation is essential. This chapter considers preparation for a formal interview because the points raised are applicable in any situation. Although directed primarily to interviews for first posts in general practice, the principles apply to interviews in general

It is worth mentioning that applications for trainee posts in the vocational training scheme do not differ from others except that they can be made only to principals who are trainers in the scheme. Letters of application, CVs and interviews are the same as for any associate or non-VTS assistant post.

Background knowledge

Interviews make most people nervous, usually because of fears of the unknown, such as the personality of the interviewer or the questions that may be asked. However, it need not be an excursion into the totally unexpected because good preparation can reduce many uncertainties.

Sometimes you may know of the principal in advance because an introduction was effected through personal contacts. Often, however, you may not know who is advertising until you are contacted following your application through a box number. As soon as you know the name and address of the principal, it is useful to consult the current *Dentists Register*. Every dental school and medical library has a copy, but if you cannot manage to find one, the dean's office at the nearest dental school may give you the information you require over the telephone because it is not confidential.

The register records when and where a dentist qualified, so assuming he or she qualified at approximately 22 years of age, as do most, you can estimate the age of the principal who will be interviewing you. Possibly a member of your dental school's staff qualified at the same school and at approximately the same time, and might be aware of the principal's particular dental or general interests. Certainly, knowledge of where he or she qualified will prevent you making the *faux pas* of the candidate who, when asked why he studied at dental school Y instead of dental school X, replied that dental school X was hopeless; an unfortunate remark because the principal had qualified there.

The register also records any additional dental qualifications, such as FDS, MDS, DDS, DDPH, DOrth or MGDS. If you do not know, find out what these letters mean and the work involved in obtaining them. For example, the MGDS is a postgraduate qualification for general dental practitioners. The studies involved demand considerable time and effort, which are, of course, in addition to the practitioner's normal daily practice workload. As the qualification is neither necessary in order to be a general practitioner, nor one which brings additional financial rewards, it indicates a principal dedicated to continuing education.

Not only is this a useful talking point at interview, but also your awareness that the principal possesses an MGDS shows that you have done your homework. Studies of industrial managers have demonstrated that interviewers are impressed by candidates who take the trouble to become fully versed about a company prior to the interview.

Self-appraisal

Interviews involve questions. Some of these can be anticipated as routine, but list as many as possible and prepare your answers. Interviews are different from viva-voce examinations in that they probe opinions and attitudes rather than basic knowledge. The following will give you the flavour of the type of question, and is a useful exercise for self-assessment, which you should perform from time to time throughout your life.

● Where do you see yourself (or your career) in 5 or 10 years' time?

● What are you good at? For example, talking to patients, crowns, oral surgery. Do not be modest—if you think you are competent, say so.

● What are you weak at? Perhaps root treatments, orthodontics. Be honest—everyone has weak areas and a good principal would be prepared to help you to improve them.

● What do you most and least enjoy doing in dentistry? These are not the same as those activities at which you think you are good or weak.

● What would you hope to gain from the practice?

● What do you feel you could offer the practice?

● How long would you envisage staying at the practice? Although you stated a time in your letter of application, you should be prepared to expand on this.

For other ideas, ask yourself 'If I were going to employ someone, and risking the reputation of my practice, what would I want to know about him or her?'.

If possible, ask someone to give you a mock interview. We all have unconscious habits, for example, too-frequent use of phrases such as 'right' or 'you know', which a mock interview will highlight. Ideally, record it on tape (or video if you have access to one) and study the recordings at leisure. You may be surprised how you sound and/or look. Who might you ask? Best, probably, would be a general dental practitioner whom you know, or one who is on the staff of your dental school. If this is not possible, anyone who is used to interviewing people or who has experienced several interviews. But be careful. Do not use a mock interview to try to change your personality, but only to identify habits that might irritate an interviewer.

As part of your preparation, you should learn the various methods of remuneration which apply in general dental practice, including that of the vocational training scheme, so that you will understand any financial proposals made by the principal. Here too, a dental school's general dental practitioners can help, as will the BDA, which would send you the relevant literature.

Getting to the interview

Personal directions can be difficult to follow. 'Turn left at the small shop and then next right after a green farm gate' can lead to disaster, especially if it is dark and raining. Buy a decent map and/or an 'A to Z' of the town, and work out how to get there for yourself. You must take this trouble because there is no excuse for being late at interview. If your 20-year-old car breaks down on the M6 *en route*, this is no excuse. You should have anticipated that it might have. If you use public transport, assume that the train/bus will be at least 3 hours late. Allow for the unexpected, and if you think that you are timing the journey too finely for a morning interview, travel the night before. Regard the cost of bed and breakfast as an investment in your future.

Try to arrive in the practice vicinity about an hour before the interview and use the time to absorb the atmosphere of the locality and to assess the patient catchment area. Study the outside of the practice because this may reflect the way it is run. Is it attractive, well maintained and professional looking or, for instance, is the telephone number displayed in a window on a felt-tip written card which would be more appropriate for a sauna and massage parlour? Study the names on the professional plates. Where did practitioners other than the principal qualify? Do any of them possess additional qualifications? There is no need to examine nearness to public transport, parking facilities and so on—you are going for an interview, not to buy the practice.

Personal appearance

Fashions change and probably the viva-suit has gone forever, but there is a world of difference between dressing in a tidy manner and looking as though you have just popped in from the beach. Whatever you wear should be neat, clean and polished, and your hair should be well groomed, whatever its length, style or colour(s). Presenting your best is, basically, a sign of respect for your interviewer. Dental surgeons are not impressed by candidates who present with dirty finger nails (yes, it happens), or if they have to sit looking at the dirty, cracked-varnished toe nails of a woman applicant wearing open-toe shoes (yes, that too).

First impressions are the strongest. It can take a lot of interview time for you to neutralise the effect of a bad first impression, so take all possible steps to present a good one and avoid opening the interview by scoring an own-goal.

Presenting for the interview

A dental practice operates as a very close knit team, and although it will (or should) be the principal who interviews you, others in the team may, unobtrusively, also be assessing you. On at least one occasion, a principal confronted with a decision about two candidates of equal merit selected the one favoured by the trusted, long-serving receptionist. The candidate rejected was the one who, on arrival, spoke to and treated her as if she were a domestic servant. You are hoping to become a member of a team, so be pleasant and courteous to everyone.

The only correct way of addressing the receptionist, or whoever you first meet other than the principal, is 'Good morning/ afternoon. My name is Joseph Bloggs and I have an appointment with Mr X. at . . .'. In particular, look at the person to whom you are speaking and smile. Avoid jocular familiarity. A candidate attending for interview with a practitioner named 'Mr D.' addressed the receptionist with the words 'Hi gorgeous, what do they call you?' and received the acidic reply 'They call me Mrs D'.

The interview

Your greeting to the interviewer should be 'How do you do', not 'Hi there' or other such phrases of familiarity. Again, look at the principal and smile. It is probable that a handshake will be exchanged. Do yourself a favour; ask a friend to shake hands with you and to tell you how it feels. Nothing is more off-putting than shaking a hand that feels like a dead jellyfish, or which is so powerful that the handshake falls just short of an assault. Practise a handshake that is gently firm.

The principal will hope to feel at ease with you, because he or she will want to employ someone with whom the patients will also feel at ease. A patient's choice of dentist is probably based more on a friendly, reassuring and caring attitude than on clinical ability. A principal finding it difficult to empathise with you might doubt your talents in this very important area.

This does not mean that you need to stare at the principal continuously, but nor should you engage in an exercise of optical isolation. I recall one interview where I wondered if there were a damp patch on the surgery ceiling because the candidate gazed at it the whole time. As a general rule, look at the principal when you start speaking, and if you look elsewhere or gesticulate when making a point, be certain to look back into his or her eyes when you finish speaking.

When speaking, make your remarks informatively sufficient. Do not reply with just 'Yes' or 'No', or, like Tennyson's babbling brook, go on and on forever. Some candidates make it virtually impossible for the interviewer to get a word in edgeways. Be alert to the principal's expression, and when you see that he or she wants to break into the conversation, pause to allow him or her to do so.

An informative CV will offer the principal a choice of topics for discussion, but as the most relevant will be your experience as a dental student, it is common to be questioned about that period. If so, never make derogatory remarks about either your school or its staff. This is not self-interested advice on my part. We have all experienced a tradesman who explains in detail that a previous workman was an incompetent halfwit. It never impresses. If you have nothing good to say about someone or something, it is better, and usually safer, to keep quiet.

It is respectful to refer to your former (or current) teachers as 'Mr X., Miss Y., Dr B.' rather than by their first names or nicknames such as 'Ham-fisted Henry'. Quite apart from sounding impertinent, and it is more common in interviews than you might think, the principal could be a relative or personal friend of 'Henry's', and has been on occasions.

At some time you will be asked if you wish to ask questions. Try not to phrase them in a way that sounds dictatorial. For instance, instead of saying 'Who would be my DSA?' use the more diplomatic question 'Would I usually work with the same DSA?'. Do not start with questions relating to your material comforts, such as 'How much holiday will I get?', or 'When will I get paid?'. If these issues have not already been explained by the principal, leave them until the end. Of course they are important, but should not appear to be your primary consideration. There are many questions which you could ask and which are unlikely to cause offence. Compile a list and consider them under the following headings: the working of the practice, local professional activities, local factors in general. Here are a few examples.

● Is the same dental laboratory used for all work, or are particular ones used for bridges and so on? How often are the collections/deliveries and do these allow a 'same-day' service for relines and additions?

● If the practice employs a hygienist, is all relevant work referred to her or only the more prolonged treatments?

● What are the GA arrangements? Are they undertaken? Is there a visiting anaesthetist(s) or are they performed by members of the practice?

● Which impression/restorative materials are used and is there a practice policy?

● How helpful are the local doctors?

● Where and who are the local dental consultants? What types of treatment are usually referred by the practice?

● Is there an active postgraduate scene locally? Where are the nearest BDA meetings held? Is there a local postgraduate centre? What are the principal's feelings about time taken off to attend courses? Is there a practice limit to the time taken per year? Are there any local dental societies?

● What is the accommodation situation locally? Are flats/houses easy/difficult to rent/buy? If you have children, what are the local educational facilities?

It has been suggested that a written list of such questions should be taken into the interview. Do so if you wish, but be wary because you could give the impression that you are interviewing the principal rather than vice versa, and on occasions candidates have offended principals for this reason. Try to memorise a few key questions from each of the groups mentioned and rehearse them during the journey to the interview.

It is assumed that during the interview you will be shown around the practice, but it is not for this chapter to discuss which features you should note. In any case, it is an easy part of the interview because the things you observe will prompt questions and topics for discussion.

Nor is it the purpose of this chapter to outline the steps to be taken if you are offered a post, other than to advise you not to make an instant decision. First seek expert advice. The defence societies and the BDA publish advice sheets and booklets which have been compiled by professionals—study them. Do not make a rushed decision just because you are anxious lest someone else might be appointed. If anything gives you reason for concern, it will the other applicants, and they too, if they have any sense, will not decide until receiving expert advice.

The exception is an interview for a trainee post in the vocational training scheme. Here, there is a formal contract between the trainee and trainer which has been developed and approved by the BDA and the DH. Therefore, if the post were being offered within the vocational training scheme, it would be reasonable to accept immediately if you wished. Indeed, one of the main advantages of working as a trainee in the scheme is the protection from unsuitable terms of service which it offers the newly qualified at this critical stage in their careers.

When the interview is over, part with words to the effect, 'Goodbye, thank you for inviting me for interview'. Expressions such as 'Cheers' or 'See you' are better confined to social occasions.

Do not worry if, on reflection, you think of things you should have said or questions you should have asked. It happens with even the most experienced interviewees, and after every interview and however well they think they were prepared. Be concerned if the interview was punctuated by periods of total silence because this could suggest that your preparation had been inadequate, and in need of revision for next time.

Conclusion

It might be argued that if everyone followed guidelines as discussed here, all candidates would be stereotyped clones. This is not so. Personality, achievements or any other of the many facets that make up the whole person will not be altered. These chapters aim to present a framework upon which to hook these facets so that the germane are neither overlooked nor presented other than in your best interest.

3

The Trainee's Contract

M. Paulson

Since January 1, 1988, general dental practice has had its own national vocational training programme. This chapter describes its administration and the trainee's contract.

The aims and objectives of the scheme were identified by the Committee on Vocational Training as follows.

'At the end of the vocational training period, as a result of working within a sheltered environment with an appointed trainer and participating in an educational course, trainees should be better able:
● to provide unsupervised and with confidence a full range ot general treatment and care to patients;
● to be aware of their clinical limitations and to refer patients for specialist opinion and treatment when necessary;
● to undertake the management skills necessary for the practice of dentistry;
● to understand the organisation of the National Health Service;
● to understand the legal and ethical aspects of the practice of dentistry;
● to be self-critical and be conscious of the responsibility to apply new knowledge to practice;
● to understand that professional training and education should be a continuing process.'
Just what have the trainers and trainees let themselves in for?

Standard contract
Every trainer and trainee will have to use a standard contract (fig. 1). The contract was drawn up by the British Dental Association, with advice from the Association's solicitors, and approved by the national committees on vocational training, and by the defence bodies.

All too few dentists have written contracts, but this contract clearly spells out the obligations which the trainee has towards his or her trainer, and also the responsibilities that the trainer owes towards the trainee. The essence of their relationship is that the trainer is the trainee's employer. As the trainee's employer, the trainer is vicariously liable for the trainee's acts and omissions. At a service committee hearing, in a civil suit, or at a General Dental Council hearing arising from the trainee's actions, the trainer may be held responsible. However, the trainee would almost certainly be called as a witness in a service committee case, could be named in a writ, and could appear before the GDC as a fully registered practitioner. As a qualified dentist, the trainee will be expected to be able to perform certain duties with the minimum of supervision, and this would have had to be taken into account at any disciplinary proceedings.

Going through the contract, the preamble relates the trainer's and the trainee's status as registered dental surgeons. The opening clause says that the trainee is an assistant, which defines a trainee as far as NHS Regulations are concerned, and that the trainee is to be the full-time employee of the trainer. For the purpose of this contract, 'full-time' is taken to mean an average of 35 hours a week, or 28 hours when the trainee is attending a day-release course during university term time.

The contract lasts for a maximum of one year, the duration of a training scheme, and while the trainee is employed as such, he or she will receive a salary set at half of target average net income (TANI). TANI is normally reviewed annually, so during the currency of a contract the salary would probably be altered, which is why the contract does not specify the amount. At the moment, the trainee's salary is £13 457. It is for the trainer and the trainee to agree when the salary is paid, as the date will vary from practice to practice, according to their different scheduling dates. The salary will be reimbursed in full to the trainer, monthly in arrears. The trainee's salary is superannuable under the NHS Superannuation Scheme, with contributions at 6%. Trainers should not have to forward the contribution to the superannuation fund, as FPCs and Health Boards may make the deduction and deal with the NHS superannuation division directly. The trainer is responsible for the employer's National Insurance contributions (also fully reimbursed) and for deducting the employee's contribution from the trainee's pay. NI is paid at contracted out rates. The trainee's salary is taxed under PAYE.

All practitioners should protect themselves against professional risks by means of indemnity insurance, and the three recognised defence societies referred to in Clause 5 are the Medical Defence Union, the Medical Protection Society and the Medical and Dental Defence Union of Scotland. The societies were of great assistance in drawing up the standard contract, which has been approved by them.

The trainer has to offer proper educational support under Clause 6, which includes allowing and requiring the trainee to attend the scheme's day-release course. As the trainer takes responsibility for the trainee's actions, Clause 7 says that the trainee must obey the trainer's directions. This is central to the master-apprentice relationship that a trainer and trainee have.

As it is the NHS's vocational training scheme, it is expected that trainees will do a full range of NHS treatment. If any private work is done, the fee will accrue to the trainer, who is still responsible as the employer for the trainee's acts and omissions.

In relation to NHS patients, under Clause 7 (viii) the trainee has to comply 'in so far as possible' with the terms of service for GDPs. NHS contractual requirements are binding on dentists who are in contract with the NHS, but as an assistant, a trainee is not on a dental list and cannot answer before a service committee for a breach of the terms of service. From the trainer's point of view, however, it is entirely reasonable that the trainee behaves as if the terms of service applied, as indeed they will do when the trainee eventually

THIS AGREEMENT is made the _____ day of
_____ 19 _____

BETWEEN _____ of _____
in the County of _____ Dental Surgeon
(hereinafter called "the Trainer") of the one part and _____
_____ of _____
in the County of _____ Dental Surgeon
(hereinafter called "the Trainee") of the other part.
WHEREAS the parties are both duly qualified and registered Dental Surgeons, the Trainer being in General Dental
Practice at _____

AND WHEREAS the Trainer having been approved as a Trainer in General Practice and the Trainee being desirous of becoming a Trainee in General Practice. AND WHEREAS the Trainee is desirous of entering employment on a vocational training programme with the Trainer AND WHEREAS the parties hereto agree to the establishment of this contract upon the terms and conditions hereinafter mentioned.

NOW IT IS HEREBY AGREED as follows:

1. THE Trainer will employ the Trainee and the Trainee will serve the Trainer as his whole time assistant in his said practice to the best of his knowledge and ability and will do his best to promote the interests of the Trainer and to serve the patients of the practice.

2. SUBJECT as hereinafter provided the employment hereunder shall last for a period of one year commencing on the   day of   19
The employment may be terminated by either party giving one month's notice in writing to the other and such notice may be given at any time.

3. DURING the continuance of this employment the Trainer shall pay to the Trainee a salary at the rates laid down from time to time in the Statement of Dental Remuneration payable to General Dental Practitioners under the National Health Service. Payments will be made by monthly instalments on the   day of each calendar month.

4. THE Trainee will be subject to the NHS Superannuation Regulations and the Trainer will deduct from the Trainee's salary and account to the proper authority for all contributions and other payments for which the Trainee is liable under the said Regulations.

5. DURING the period of this employment both parties shall become and remain at their own expense members of a recognised medical defence organisation.

6. DURING the continuance of the employment the Trainer shall:
(i) be available to the Trainee for guidance in both clinical and administrative matters;
(ii) provide adequate bench reference books for the use of the Trainee;
(iii) allow and require the Trainee to attend the appropriate day release course of approximately 30 days in the year arranged by the Regional Adviser/Associate Adviser in General Dental Practice;
(iv) provide the Trainee with satisfactory facilities so that a wide range of NHS practice is experienced and so that as far as is reasonably possible the Trainee is fully occupied;
(v) provide the Trainee with adequate administrative support and the assistance of a Dental Surgery Assistant; and
(vi) inform the Regional Postgraduate Dental Dean if the circumstances of either the Trainer or the Trainee change in such a way as to alter the contract of employment.

7. DURING the continuance of his employment the Trainee shall:
(i) fulfil and obey all lawful directions and orders of the Trainer from time to time and not at any time except in the case of illness or other unavoidable cause or permitted holidays absent himself from the service of the Trainer without the Trainer's consent;
(ii) keep proper accounts of all professional visits fees paid and all patients attended operations performed and prosthetic work and all other business done by him for the Trainer and of all monies he shall receive or pay on the Trainer's account and forthwith pay all monies so received to the Trainer or as he may direct. The Trainee shall keep all usual and necessary dental charts and an approved register of the work done for all patients attended to by him;
(iii) devote his whole time to the practice of the Trainer during the usual business or professional hours. Such hours which may include the provision of out of hours services shall be agreed in advance and a note thereof shall be appended to the foot of this Agreement;
(iv) not without the written consent of the Trainer be employed in any way or for any purpose by any person company or firm outside the usual professional hours of the practice;

(v) neither without first obtaining the written consent of the Trainer attend any patient or perform any operation or prosthetic work for any person other than the Trainer nor on his own account either carry on or be engaged in a dental practice nor accept any part time or full time dental appointment whether paid or unpaid or give any dental advice either gratuitously or for reward;
(vi) not whether during or after his employment disclose any professional secrets or any information with respect to the Trainer or his family patients practice or affairs or any directions given to him by the Trainer;
(vii) observe and conform to the provisions of The Dentists Act 1984 so far as they relate to him or his employment and observe and conform to all the laws and customs of the dental profession;
(viii) in relation to any patient treated by him or desiring to be treated by him under the National Health Service comply insofar as possible with the Terms of Service applicable to dentists under the provisions of the National Health Service (General Dental Services) Regulations;
(ix) attend such day release courses as are set out in the published programme and shall not except in case of illness or other unavoidable cause or permitted holidays absent himself from any such course without the previous consent both of the Trainer and of the Regional Adviser/Associate Adviser in General Dental Practice;
(x) undertake such educational studies as may be reasonably advised from time to time by the Regional Adviser/Associate Adviser in General Dental Practice;
(xi) inform the Regional Postgraduate Dental Dean of any alteration in his circumstances which might affect this contract of employment.

8. THE Trainee shall be entitled with full pay to four weeks' holiday during the period of twelve months in the practice and pro rata for shorter periods. Such holidays shall be taken at the times agreed between the parties. In addition the Trainee shall be entitled to public holidays.

9. IF the Trainee is absent due to sickness or maternity the Trainer will pay over to the Trainee such sums as the Trainer may receive for the Trainee's salary in accordance with the Statement of Dental Remuneration.

10. A period of absence may be extended beyond four weeks only with the consent of the Trainer and the Regional Adviser/Associate Adviser in General Dental Practice. Such extension shall not exceed a further four weeks. If permission for an extension of leave is not given, the contract shall be terminated forthwith by the Trainer.

11. FOR a period of one year following the completion of the employment the Trainee, unless practising in the Trainer's practice or otherwise with the Trainer's consent, shall not:
(i) accept as a patient any person who was during the training programme a patient of the Trainer or one of his partners associates or assistants;
(ii) attend or treat in the capacity of a General Dental Practitioner any such patient as is mentioned in (i) above;
(iii) recommend or induce any such patient to seek treatment from any dental practitioner other than the Trainer and his partners associates or assistants;
(iv) The provisions of subclauses (i) and (ii) above shall not apply in the case of a patient seeking emergency treatment.
Each of the subclauses (i) (ii) and (iii) shall be separately enforcible as if each were independent covenants.

12. NOTHING herein shall entitle the Trainee to any of the rights or expose him to any of the liabilities of a partner or constitute in any way the relationship of partners between the Trainer and the Trainee.

13. ANY dispute between the parties or those in any way representing them concerning this agreement or the employment of the Trainee or anything arising from this agreement shall be referred under the provisions of the Arbitration Acts 1950 and 1979 to a sole arbitrator nominated by the Secretary of The British Dental Association.

PTO ▶

14. IN this agreement words denoting the male gender shall include the female gender and references to any enactment order regulation or other similar instrument shall be construed as a reference to such enactment order regulation or instrument as amended from time to time or as replaced by any subsequent enactment order regulation or instrument.

AS WITNESS the hands of the parties hereto this day and year first before written.

SIGNED by the said Trainer:

in the presence of:

SIGNED by the said Trainee:

in the presence of:

The usual practice hours are:
The out of hours services to be provided by the Trainee are:

Fig. 1 The contract, available from the BDA, 64 Wimpole Street, London W1M 8AL. It comes in a folder which carries notes of guidance.

becomes an independent practitioner.

Clauses 8, 9 and 10 deal with leave entitlement. If a trainee agrees to work on a public holiday, he or she should be given a day's leave in lieu. Clauses 9 and 10 entitle the trainee to take 4 weeks' sick leave; they also allow the trainee 4 weeks' maternity absence (to be counted separately from sick leave) with the trainer's permission. Any extension of leave has to be subject to the trainer's and the course organiser's approval, without which the contract is immediately terminated.

Binding out

The binding out clause in the contract (Clause 11) has attracted a lot of attention. The clause is designed to protect the trainer's practice after the trainee's departure, and can only be waived by the trainer. The reference to 'assistants' in subclause (i) includes the trainee. In other words, the trainee should not subsequently accept as a patient someone he or she treated at the training practice. The Association's legal advice was that the conventional binding out clause would not be enforceable against a junior dentist who was in training. In the context of the training contract, by placing a mileage restriction on the trainee's freedom to practise, the trainer would be attempting to set up a monopoly which could be considered to be against the public interest. Whilst the clause allows the trainee to set up next door to the training practice, its terms still comprehensively protect the existing patients and goodwill by the practice.

The standard contract, with accompanying notes of guidance, is available from postgraduate deans and course organisers, and from the British Dental Association, to members and non-members.

Terminology

The notes of guidance make it clear that different areas use different terminologies. Thus, in Scotland the postgraduate dean is known as the regional general practice vocational training adviser (RGPVT adviser) and a regional adviser is known as a general practice adviser (GP adviser). In Northern Ireland the postgraduate dean is called the postgraduate tutor. Similarly, the equivalent to the Family Practitioner Committee is, in Scotland, the Health Board, and, in Northern Ireland, the Central Services Agency.

FP17s

As an assistant, the trainee will sign estimates 'pp' the trainer's name, and under the trainer's stamp, but there will be a suffix to the trainer's contract number to enable the DPB to identity work done by trainees, for statistical purposes. This suffix will be two digits: a 5 followed by a 1 for a trainer's first trainee, 2 for the second, and so on. As the scheme starts, one or two associates may become trainees, and will need to submit two forms for treatment which is uncompleted on the transition date: one for the first part of the treatment, under the associate's own number, the second for its completion, under the trainer's number plus suffix. Where the trainer and the trainee share a course of treatment, separate FP17s should be used for the work they individually carry out.

Trainer's contract

As well as entering into a contract with the trainee, the trainer also has to agree to certain terms in return for approval as a trainer. The terms are given in a letter from the postgraduate dean (fig. 2). The letter makes it clear that the trainer's grant, which is set at 15% of TANI, and is currently £4037, will only be payable when a trainee is in post. The grant is superannuable.

Becoming a trainer

Details of training schemes are circulated to all GDPs within each region by the postgraduate dental dean (RGPVT adviser in Scotland, and postgraduate tutor in Northern Ireland). Dentists wishing to become trainers have to complete an application form which will be available from the dean's office. A trainer should be an experienced dentist with high clinical and ethical standards, and who provides a wide range of treatment. Ideally, he or she should have managerial responsibilities within a practice.

Applicants are interviewed by a selection committee, convened for that purpose by the Regional Postgraduate Dental Education Committee (except in Northern Ireland where it will be a subcommittee of the CPME). In England and Wales the selection committee should include the dean, the regional adviser, a GDP from the RPDEC, a GDP nominated by the Royal College of Surgeons, a representative of the GDSC, and a representative of the chief dental officer. The selection committee can also seek references. Upon receipt of a completed application form, arrangements would be made for a visit to the practice, usually by the regional adviser, prior to the interview.

The training practice should be adequately equipped and should have modern sterilising and easily accessible radiograph equipment, and the physical layout of the practice should be such as to ensure that support to the trainee can be provided by the trainer for not less than 3 days a week.

Approval as a trainer lasts for the duration of a scheme (one year) and the decision of the selection committee is final. Rejection does not prohibit future applications, however, nor does rejection by one region prevent a dentist applying to another. Exceptionally, joint trainers within a practice may be appointed.

Becoming a trainee

It is expected that trainees will be recent graduates. The appointment of a trainee is entirely a matter for the trainer. Deans will advertise their schemes and circulate brief details of approved practices to prospective trainees, who make their own arrangements for interviews.

Once a trainee is appointed, it is up to the trainer to tell the FPC, Health Board or CSA of the appointment. A copy of the contract between the trainer and trainee has to be deposited with the dean.

Dear . . .

Vocational Training Scheme for Dental Practice—VT Scheme 19

Thank you for attending for interview. I am pleased to inform you that the Selection Committee which met on . . . has recommended your approval as a trainer in dental practice to the Regional Postgraduate Dental Education Committee, subject to the appointment of a trainee in your practice. S/he will take part in the vocational training scheme based at . . . during 19 . . .

As an approved Trainer you will receive a training grant (15% of TANI currently £4037 p.a.) from your FPC and your trainee will receive a salary (50% of TANI currently £13 457 p.a.) which will be paid to him/her by you and reimbursed by your FPC along with your employer's National Insurance contributions.

Please note that your approval by the RPDEC as a trainer in general dental practice is related only to the 19 . . . vocational training scheme, and that the payments due to you as an approved trainer (see above) are dependent upon the appointment of a trainee in your practice. The training grant will be paid by your FPC monthly in arrears while your trainee is employed in your practice. You will receive a maximum of 12 monthly payments. The grant will be in recognition of your acceptance of the responsibilities of a trainer. The detailed responsibilities are set out in the nationally agreed contract (copy enclosed) but the main provisions may be summarised as follows:

● to have in your practice at (address) a vocational trainee in full-time employment;
● to act as trainer to the trainee during his/her time in the practice, working in an adjacent surgery for not less than 3 days per week, and to offer guidance in both clinical and administrative matters;
● to ensure that the work experience and training log provided for the trainee is maintained throughout the year;
● to allow and require the trainee to attend the day-release course of 30 days, held on . . . days during 3 university terms (approximately 10 days a term) starting on . . . 19 . . . and to ensure that the trainee takes holidays outside term times; absence from the day-release course for reasons other than sickness will only be allowed in exceptional circumstances and after written application in advance to the adviser in general dental practice, Mr A. N. Other; failure by the trainee to comply with this requirement will be interpreted as withdrawal from the scheme and will in consequence affect your own position as an approved trainer;
● to employ the trainee as a salaried assistant under the nationally agreed contract and before the trainee starts work to deposit a copy of the signed contract with the postgraduate dental dean (any intended variations in the standard contract should be notified to the postgraduate dental dean in advance);
● to provide the trainee with satisfactory facilities so that a wide range of NHS practice is experienced and so that s/he is fully occupied in clinical practice;
● to take part in not more than 14 sessions of appropriately organised educational sessions during the year as laid down by the postgraduate dental dean; this includes attendances at the day-release course for not less than one half-day per term (the dates of the trainer study days will be notified to you by the adviser in general dental practice, A. N. Other);
● to inform the dean of any alteration in the circumstances of the practice, the trainee or yourself which might affect your approval as a trainer. You will appreciate that inability to meet the responsibilities stated above for whatever reason would preclude your continuation as a trainer during the 19 . . . scheme.

I would be grateful for your acknowledgement that you are willing to undertake the responsibilities of a trainer, by signing and returning the attached slip.

I look forward to hearing from you.

Yours sincerely,

Name and title of regional postgraduate dental dean

Trainer's name and address

UNIVERSITY OF . . . .
Vocational Training Scheme for Dental Practice — . . . . 19..

Name ....

Address ....

I accept the responsibilities of a trainer on the 19. . . . . . vocational training scheme for dental practice as specified in the contract letter dated (date).

Signed .... Date ....

Please complete this slip and return it to: Postgraduate Dental Dean, etc.

Fig. 2 Terms of approval as a trainer—model letter from the postgraduate dental dean.

4

Recruiting Practice Staff

J. Muir

Taking on employees is not the simple business that it may seem. Dentists do not necessarily have the weight of experience of recruitment to ensure that they employ the best receptionist, dental hygienist or auxiliary worker and that the appropriate paperwork is completed to comply with employment law. This chapter considers job descriptions, interview questions, references and promoting the practice as an employer.

Sound recruitment is the best foundation for an effective working relationship. Thereafter the need on the part of the employer to ensure that standards are met should suggest an awareness of rights and responsibilities under the contract of employment; and if standards are not met, an appreciation of the course open to the dentist to bring the contract to an end by dismissal. While the demands of patient care will invariably be high, the dentist, as employer, cannot afford to overlook the fact that poor recruitment and inadequate attention to employment practice can result in substantial penalties. Poor performance on the part of staff will undermine the economic viability of the practice and a badly handled dismissal may well lead to the payment of substantial compensation to the ex-employee on the grounds that it was unfair.

The job

If a post falls vacant or one is to be created, it is essential to decide what actually needs to be done in the post and to highlight the main points in a job description. This need not be an elaborate affair, but from the practice's point of view it will clarify the parameters of the job and from the candidates' standpoint it will suggest what they need to show as regards experience and background when applying for the post. A focus of this kind is extremely useful in getting the business going.

The practice may not have an application form but if there is one, does it ask the right questions, is too little information sought and, just as a practical point, is there enough space? There is much to be said for candidates applying for posts with their own written submission. This way the employer can make some useful observations on the way the application has been put forward. Is there clarity of thought, or does the application go on and on, perhaps indicating a trait of long-windedness?

The interview

On the basis of written applications, a dentist can draw up an interview shortlist. Again, despite the demands of patient care, the dentist should avoid a situation where interviews are squeezed in and, in all probability, rushed between patients' treatment. Time should be set aside. The dentist should consider the questions that need to be asked so as to explore fully with the candidate how previous experience, educational background, training and so on fit in with the post in question. All too often an interview session of this kind is completely unstructured. Vital areas of discussion are missed out and there is no viable basis of comparison between one candidate and another. Questions can be badly phrased in that the answers can be 'yes' or 'no'. This does not permit candidates to put their views forward and certainly does not enable the interviewer/employer to draw useful conclusions. Again, care is needed to ask questions of an open-ended character so that the candidate does the talking.

Obviously interview time is going to be limited. One major failing on the part of many potential employers is that they spend a disproportionate amount of the time available talking about the firm or the practice and, at worst, about themselves. The whole object of the exercise is to put the candidate in the spotlight and, through friendly and constructive questioning, to find out how he/she can fill the post. It is to be stressed that interview questions should not be framed aggressively. The interview is not a battle of wits. If the questions are provocative, it is to provoke response, not to alienate candidates.

Another important consideration is note-taking. Perhaps interviews have been held over a period of several days so as to suit the convenience of practice demands as well as candidates' availability. At the end of a busy day it can easily be a question of 'which candidate was that?' Some people are able to carry impressions and information in their heads but experience suggests that memory dims quite quickly. If interviews have had to be spread over a few days, early candidates, in the absence of refresher notes, are not necessarily going to fare well. It should also be borne in mind that objective assessment is a sound defence against complaints of discrimination, and notes which support that procedure may be vitally important material.

There is no reason why notes should not be taken during the interview, although it is best to tell the applicant this at the outset. However, if note-taking affects the flow of business, and the dentist may have this in mind more so than the candidate, then reserve the note-taking until after the candidate has left. The clear advice is to get on with the job straightaway.

Employment history

There are a number of areas that need to be checked out with applicants. All applicants have a history. Much of this will emerge in the process of a good interview. If there are gaps, how did they come about? It is vitally important for the employer to know. While the great majority of applicants are truthful about previous jobs and work experience, there are advantages in calling for references. It is sometimes difficult for the applicant to get his/her current employer to surface with a reference on the simple basis that if the job does not materialise then the applicant has shown his hand, but a reference subject to a job offer is perfectly in order.

Timing is all important. If a job offer is made and accepted but the reference is not called for until later, then major difficulties could arise if the reference is found to be unsatisfactory, not the least of which is a problem for the dentist. Does he/she continue with an individual who is now known to have a cloud over him/her or does he scrap the whole arrangement and start again? Clearly the employer is entitled to withdraw, but the problem need never arise if everything is done in proper sequence.

Dentists must also consider any medical condition candidates may have. The practice requires staff who are going to be perfectly normal attenders. It would seem to be a heavyweight process to require a medical examination before engagement, but the dentist should seek assurances about health and make an objective judgement about the response against the requirements of the job.

Towards the end of the interview an opportunity must be afforded to the candidate to ask his/her questions about the post and the practice. The main attractions of the job, that is the essential terms of employment, will presumably have been spelled out in the original advertisement, or to the recruitment agency for passing on to interested applicants. A candidate may, however, wish to explore whether leave increases with service, whether there is bereavement leave, what happens about overtime payments, whether pay is geared to an NHS grade and whether there will be a review of pay after 6 months or annually.

This is the opportunity for the dentist to outline the practice policy. It presupposes that policy matters of this kind have been thought about, so for any practice going in for recruitment it is as well that such issues are considered in advance so that appropriate answers can be given. Uncertain or 'it depends' answers cut no ice with prospective employees. Candidates may well be content that not every question can be answered there and then but they will expect enough positive and reasonably-based answers to suggest that the practice knows what it is doing.

From the responses the dentist might make to specific questions and the general information that has been supplied about the practice and its future, applicants will no doubt be forming their own opinion about the practice as an employer. Selection is a two-way process. The applicants may be seeking a better job than they currently hold and are therefore very interested parties. The fact is that applicants are also selecting the employer to an extent. There is inequality when there is a massive response to a job advertisement, in areas of high unemployment for example, but in the reverse situation, it is as much a question of applicants turning down the practice as the practice rejecting applicants. This should suggest that a practice is put on its mettle in advertising itself as a would-be employer.

A parallel task to the initial exercise of drawing up a job description is to consider what qualities and attributes the practice needs to look for in a successful applicant. Headings might include experience, academic and professional qualifications, personality, dress and presentability, speech, power of expression, potential, confidence and ability to draft correspondence. Having determined what features are sought, the practice interviewer should then make an objective assessment of each candidate against the criteria. Depending on the type of job being filled, some features may carry more weight than others. The outcome of such an exercise will often point conclusively to the best candidate.

What part should personal compatibility play in all this? In very small practices there is no doubt that the ability to get on with people and fit into established routines are matters of great importance. In large practices with a bigger staff complement, personal qualities still rank high but care should be taken to avoid the attraction of personal likes in coming to a selection decision. The object is to find the most appropriate candidate on the basis of an anticipated professional working relationship, not a personal one. The paramount consideration is whether the new recruit in whatever capacity— gardener, receptionist, bookkeeper—is going to strengthen the practice and make it more effective.

While recruitment may seem a drag and an unwelcome diversion of resources away from patient care, it is nonetheless a vital task for the practice. Time spent on thinking through the requirements of the job and what is wanted in an ideal candidate is time properly invested. The selection activity should be conducted on a structured framework and the final decision made on the basis of objective criteria. Particularly in small organisations, and the great majority of practices would fall into this category, the cost of bad recruitment can have a telling and adverse effect on practice performance.

5

Terms of the Contract of Employment

J. Muir

The process of recruitment of staff has to be within the statutory requirements of, principally, the Race Relations Act, the Sex Discrimination Act and the Rehabilitation of Offenders Act. The terms of employment form the basis of the contractual relationship between the practice as employer and the employee. Rights, responsibilities and work are all governed by the contract. This chapter considers the statutory requirements, the contract, its specific terms and implied terms.

If a job offer is accepted, a contract of employment is entered into. A distinction has to be drawn between a contract for services and a contract of service (employment). The former applies when an individual is in business for himself/herself —an independent contractor—and enters into a contract with another to provide services, for example as a dental technician or bookkeeper. They may get their work through the practice, but will not be, as self-employed persons, employees. Since they are not employees, they will not be able to claim employment rights, for example, to bring a complaint of unfair dismissal if no more work is provided.

Many organisations engage people to do work on this basis. However, it is to be noted that the distinction between employee status and self-employed status is not always clear cut and the self-employed person may be able to establish employee status for the purposes of employment protection legislation. Industrial tribunals will look behind the ostensible relationship and examine the realities. Depending on degree of control, freedom to accept or refuse work, ability to subcontract and other tests developed by the courts, a decision will be made whether the arrangement amounts to a contract for services or an employment contract.

The contract

A contract of employment exists between each employer and employee. It does not have to be in writing. A practice making a job offer across the table—'the pay is £X, the hours of work are Y hours, and so on'—followed by acceptance— 'I'll report for work on Monday'—constitutes a legally binding agreement. There are some special employments, such as apprenticeships, or seamen on UK registered ships, where the contract has to be in writing, but otherwise there is no requirement. Many hundreds of thousands of employments proceed on the basis of no paperwork.

The difficulty comes at some later stage, in determining precisely what terms are in the contract. In the first place, what was said across the table is likely to have covered only basic information such as pay, hours, holidays and overtime. It may not have been made clear whether the overtime was voluntary or compulsory for example, so a specific problem later arises. Set down the terms and conditions of an employment contract in writing and express them clearly. There is then far less scope for disputes about the exact nature of the job and the benefits and obligations that go with it.

Implied terms

Apart from the specific terms in the contract, there are also implied terms such as the employer's duty to pay for work done, to act reasonably and to be bound by the contract. On the employee's side, there is a duty to cooperate, obey reasonable instructions, to act with fidelity and so on. These implied terms can be of the greatest importance in future working relationships.

If the employee breaks his side of the bargain, the employer will have disciplinary grounds for moving against him. If the employer acts unreasonably, for example by harassing an employee, or fails to honour a prime term in the contract, for example by withdrawing use of a company vehicle for domestic purposes, the employee may have grounds for resigning and claiming constructive dismissal. The final decision on whether the dismissal was fair or unfair is another matter.

The law

Identifying the terms and conditions in a contract is still a problem, given that many employments are without written terms. The law intervened in this situation but the problem persists, despite the fact that since 1963, and the then Contracts of Employment Act, employers have been under a duty to set down in writing certain particulars about the terms of the job or to refer the employee to an accessible document.

The current requirement is in Section I of the Employment Protection (Consolidation) Act 1978. It is to be noted that the Section does not call for a written contract to be produced, but for the main terms of employment to be set out within 13 weeks of the start of employment. While an S. I statement does not itself constitute a written contract, an industrial tribunal would see it as strong evidence of the contractual terms. If an employee is asked to sign to say he has received an S. I statement that does not alter the position, but if he signs to say he has accepted the terms, the document has contractual force.

The matters which the law requires the employer to set out are as follows:

● Identification of the parties. Name of the employing organisation and the employee.
● Date when employment began. Since many aspects of employment protection depend upon a service qualification and, indeed, the amount of any redundancy payment under the state scheme depends, amongst other factors, upon total length of service, the date on which an employee joined is of great importance both to him and the employer. In very many cases the date from which service clocks up is the date when the employee joined that firm, but there may be service with a previous employer which counts. Publication gives the

employee an opportunity to challenge the date if he thinks it is wrong.

● Pay. The scale or rate of remuneration and the method and intervals at which remuneration is paid have to be set out. Remuneration should be taken to include benefits such as severance pay, overtime rates and subsistence allowances.

● Hours of work. Most employments are expressed as hours per week, but the concept of annual hours of work is gaining ground. Daily attendance hours should also be described.

● Holidays and holiday pay. The number of days of holiday entitlement have to be set out, together with any service or grade additions. If, within an employment, the working week is geared to, say, 4 days' or 6 days' attendance, the holiday entitlement has to be expressed accordingly. Problems can arise over entitlement when employment comes to an end during the holiday year, so the particulars have to cover the precise pro rata arrangements which apply. If any outstanding holiday is turned to cash, the precise formula by which this is done must also be described.

● Whether the employer has a sick pay scheme. The scheme of Statutory Sick Pay is not a condition of service, since it is a transfer of function from the DH to the employer. However, it has to be spelled out whether the employer has an occupational sick pay scheme or not. If there is no employer scheme, the particulars must say so. If there is a scheme, then the details have to be set out. Because there may be a good deal of detail about entitlements, service qualifications, medical and other certificates, notification of absence and so on, it is usually more convenient to cover these features in a supporting document.

● Pension scheme. If the practice has an occupational pension scheme then details must be set out, preferably in a supporting document.

● Notice periods. There are statutory minimum periods of notice which must be observed as follows: subject to one month's employment, the employee must give one week's notice; subject to one month's employment the employer must give, up to 2 years' service, one week, then one week's notice for every year of service up to a maximum of 12 weeks' notice for 12 years' service or more. There is nothing to stop the employer from giving longer than the statutory minimum.

● Job title. The job description must be detailed enough to enable the employee to identify accurately the job he applied for and was subsequently offered. A vague description would not meet the requirements of the particulars, but would give the employer a great deal of scope and flexibility in using the employee. On the other hand, a very full description might lead to a situation where the employee was entitled to say that since so many features of the job had been described, the feature that the employer wanted to introduce and which was outside the description was not one that he was prepared to undertake.

● Disciplinary rules. If an employer wants to rely on a rule then it must be published in the contract. Such a rule might be 'smoking in the treatment rooms is strictly prohibited and breach of this rule will lead to dismissal'. It is a mistake to assume, however, that breach of disciplinary rules leading to dismissal puts the employer in the clear as regards fairness in dismissal. The industrial tribunal will no doubt examine the rule for fairness, as well as weighing up whether the employer acted reasonably in the circumstances.

● Appeal and grievance. The statement must name 'a person to whom the employee can apply if he is dissatisfied with any disciplinary decision relating to him and a person to whom the employee can apply for the purpose of seeking redress on any grievance relating to his employment and the manner in which such application should be made'. Further to these requirements is the need to point out any 'further steps consequent upon such application'.

Over a period of time it is likely that the content of conditions of service will change and the legislation requires employers to inform employees of such changes within 4 weeks.

The contract of employment is an agreement entered into freely by the parties. They can agree whatever they want, subject to any statutory requirements and the contract conditions themselves not being illegal. Thus, a contract can specify longer periods of notice, make provision for severance payments in excess of the state scheme or give maternity pay for a longer period than the statutory requirement.

The list of all the contractual features between employee and employer can be extensive. If so, the main terms and conditions (as required anyway by S. I of the EP(C) Act) should be put into the letter offering the applicant the job, which, when signed on acceptance, becomes the contract of employment. The rest of the features should be in supporting documentation.

Amendments

Contracts change, not only in obvious ways like an increase in pay from £X to £Y, but also through changes in custom and practice. A contract which calls for attendance at 8.30 am may be changed by custom by an individual who cannot get to the practice before 9 am because the bus service has been retimed. It is important to recognise that custom and practice terms have great force, and the opportunity should be taken, from time to time, to review all working arrangements and, if necessary, to redefine the contract.

The interview can also have an important bearing on the contract. If the employer gives an unrealistic picture by speaking of prospects, opportunities and responsibilities that are wide of the mark, the applicant, later the employee, may be able to rely on the statement. The company may then be found to be in breach of the contract by not honouring it.

Whether it be a statement or a written contract, it is possible that it is incomplete in the sense that not all the particulars are covered. If no statement is forthcoming, an employee has the right to go to a tribunal and get an order for one; similarly, an industrial tribunal can be asked to determine what particulars need to be included. A tribunal does not have the power to effect amendments to contractual terms outside the particulars that need to be identified in an S. I statement, but where a matter is far from clear, a precise term can be worked out.

Well-based employment on the part of a practice depends upon good recruitment, identification of all the appropriate terms and conditions of employment and putting them to the employee formally through a contract. This way it is possible for both parties to know where they stand as regards rights, duties and responsibilities.

6

Ending the Contract of Employment

J. Muir

Most employment contracts come to an end in straightforward circumstances, but redundancy and dismissal are more problematic. This chapter considers the legal implications of ending the contract of employment.

When employment contracts come to an end in ordinary straightforward circumstances, such as reaching retirement age or, on the employee's part, an unforced resignation to look for or take up another job, there is mutual consent and no question of statutory rights arises. In other situations the contract is terminated by operation of law, so, for example, the death of an employer or an employee will terminate the contract. If there is a partnership dissolution, a compulsory winding up, or the appointment of a receiver by the court, all employees' contracts come to an end.

Redundancy

Another way in which a contract ends is by dismissal on grounds of redundancy. The strict conditions are laid down in Section 81(2) of the Employment Protection (Consolidation) Act 1978:

'For the purposes of this Act an employee who is dismissed shall be taken to be dismissed by reason of redundancy if the dismissal is attributable wholly or mainly to—
(a) the fact that his employer has ceased, or intends to cease, to carry on the business for the purposes of which the employee was employed by him, or has ceased or intends to cease, to carry out that business in the place where the employee was so employed, or
(b) the fact that the requirements of that business for employees to carry out work of a particular kind, or for employees to carry out work of a particular kind in the place where he was so employed, have ceased or diminished or are expected to cease or diminish.'

Circumstances that would require the employer to act may be, for example, population drift, which reduces demand for the practice, so that staff resources are bigger than required, or a decision to transfer the accounting arrangements hitherto done by an employee in the practice to an outside agency. In any redundancy situation, it is of crucial importance to consult the employee(s) concerned, so as to see whether alternatives are available to redundancy.

Capability and conduct

Other situations where the practice may have to act and effect dismissal are related to capability and conduct. Before looking further at these two features a number of other points have to be considered.

Are the standards expected by the practice made clear to the employee or can he/she turn round and say 'I didn't know I was doing anything wrong'? Some matters can be dealt with in the contract of employment. For a receptionist an appropriate term can be written into the contract so that the employee knows that a clean, smart, well-groomed app-

earance is required. If there is no such prior requirement and, after a while, the receptionist turns up with coloured hair and faded jeans, the practice must say what standards are required and give the employee a chance to comply.

Within the contract of employment there are express terms and implied terms. The express terms cover matters such as pay, hours and holidays. The implied terms are ones that in this context are taken as read. There is no need to spell out that an employee must not be drunk at work, fight, wreck the apparatus, use abusive language, and so on. There is an implied term that the employee will do none of these things. Nevertheless, in drafting a disciplinary procedure, a number of employers would spell out such matters, with a warning that if any one of these actions is committed, dismissal will follow. It can be seen as driving home the point.

Similarly, in the matter of capability there is an implied term in the contract that the employee is able to do the job for which he/she is employed and to display reasonable competence in doing it. But in the capability field it is important to ensure that employees do know the standards to which they are required to work. If the new entrant shows requisite competence on assuming the job but over a period of time performs at a lower level, the practice will presumably be aware of this. From the employee's viewpoint, since nothing might have been said and nothing might have been written down about standards, an assumption might be made that the contribution is perfectly acceptable. It all points to the need on the employer's part to have standards and to make them known. Then the practice can measure any gap between the standard and failure to achieve it and take appropriate action The existence of standards and the objective measurement of any shortfall give the practice firm ground on which to move if disciplinary action is called for.

The law

The law specifically requires the size of the organisation to be taken into account by tribunals dealing with cases. Section 57(3) of the Employment Protection (Consolidation) Act as amended by S.6 of the Employment Act 1980 says: '. . . the determination of the question whether the dismissal was fair or unfair, having regard to the reason shown by the employer, shall depend on whether in the circumstances (including the size and administrative resources of the employer's undertaking) the employer acted reasonably or unreasonably in treating it as a sufficient reason for dismissing the employee; and the question shall be determined in accordance with equity and the substantial merits of the case'. While size and resources have therefore to be considered, there is no licence even for the smallest organisation to ignore basic natural justice factors in taking disciplinary action against an employee. It could be argued that in a very small practice an employee would know when he/she was not performing up to standard and that warnings would not be necessary. Such an argument may succeed, but it is better management practice to adopt a simple procedure.

If the job is not being done to the standard required, and this of course includes lapses in conduct as well as incapability, the employee should be told quite formally what the complaint is and be given a chance to improve. A timescale for improvement should be set and advice and help offered on how to achieve the improvement. It would be nonsense to suggest that every lapse should be dealt with on this basis. A counselling discussion, free from any overtone of disciplinary action, will often achieve the objective of getting the employee up to standard, but the dentist should not use this informal approach as a substitute for formal action. If a formal warning is ignored, the situation leads naturally to one where the dentist should give a written warning with a very clear indication that if there is no improvement then dismissal will follow. Provision should therefore be made for a final meeting before hard and fast decisions are taken on whether employment is to continue or not.

It is part of the natural justice format that any employee should be able to state his/her own case and ask a colleague to come along to any hearing. On occasion, the presence of a trade union official or representative may be unavoidable. If the practice is small, it is unlikely that there will be anyone to appeal to if a decision is made to dismiss, although again this would normally be a natural justice feature. It does, however, strongly suggest that in a large practice the senior partner should take on this role. Adherence to the principles outlined above will go a long way towards a defence that the dentist has acted reasonably in dismissing an employee.

Ill health

Within the term 'capability' is ill health. If someone turns up for work for all the required hours in the week but falls down on the job through sheer inability, then there is incapability. The same principle applies to sick absences. While the employee may turn in a thoroughly competent job while actually attending the practice, the fact remains that the employee is incapable of doing the job if he/she is not there. Most people who say they are sick are genuinely ill, though it has to be recognised that people have different thresholds when it comes to determining whether their condition is one that leads to absence. That aside, most absences are genuine. If an absence on reported grounds of ill health turns out to be malingering or a cover for some other reason, it is misconduct and should be dealt with as a disciplinary issue.

What constitutes an unacceptable level of absence due to ill health? There really is no clear measure, since employers' perceptions vary so widely. Perhaps a better way of assessing the situation is to consider what the costs are of any absence and to ask the question 'can the practice afford it?' Affording it is not just a simple question of cash out of pocket. If a practice has no occupational sick pay scheme and employees rely on statutory sick pay, then the money the employer pays out can be recouped from money the employer would have paid over to the DH and, if that is not enough, from tax due to the Inland Revenue. In these circumstances it could be argued that the practice bears no extra cost if a temporary replacement is hired. The fact remains that there is disruption and a loss of efficiency. If, however, there is an occupational sick pay scheme which pays full pay for a period and maybe half pay for a further period, the costs to a practice can be significant, apart from the disruption costs. Disruption costs are usually heaviest when absences are short but frequent, allowing no adequate steps to be taken to find a replacement.

A staff member who is often away sick also produces an adverse reaction amongst those who are left—a particularly important point where the total complement is small. There may be a feeling that he/she 'is getting away with it again' and certainly, in the longer term, a resentment that the uncovered work, in whole or in part, is farmed out to those who are left.

To help the employer be aware of the exact extent of the absences, the practice ought to have a record system. All too often a consistently high level of sick leave is taken as normal after a time, so to see what is actually happening, monitor the situation by looking regularly at the records. If the level of absence is shown to be unacceptable, action must be taken to safeguard the interests of the practice.

Action

The first step is to draw the record to the attention of the employee in writing. On the assumption that the absences are genuinely due to sickness, there is no question of warnings as would be given in disciplinary proceedings. It is often sufficient, once the record is made clear, coupled with an invitation to discuss any problems with the employer, to suggest to the employee that a more realistic effort to attend should be made. If the situation does not improve, then refer the employee to a doctor nominated by the practice. Explain to the doctor the background to the absences, the nature of the job and ask the question 'in the future is this employee likely to be a regular attender?' Ask the employee to submit his/her own evidence.

On the basis of this information it is crucial to consult the employee and see whether the problem can be solved. Perhaps shorter hours can be worked; maybe the job can be changed. The practice does not have to turn itself upside down to accommodate an employee, but it has to show that proper consideration has been given to alternative arrangements. Then if, as an employer decision, a dismissal on grounds of ill health is the appropriate course, the dentist is likely to be able to show that he acted reasonably.

Dismissal on whatever grounds is a difficult business and calls into question personal qualities on the part of the employer. Sentiment, perhaps rightly, plays a part in overlooking an employee's faults or his/her sickness record, but it ought to have a fairly limited rein. The interests of the practice should dictate that action is taken in the appropriate way; and if the issues are looked at objectively then, bearing in mind the principles set out above, the decision will be more soundly based and much easier to effect.

7

The Dental Hygienist in General Practice

M. R. N. Collins

Dental hygienists have become very valuable members of the dental team. Their training is thorough and work carried out by them should be of a high standard. It is essential that prescribing dentists fully understand the regulations with regard to dental hygienists and the responsibilities which have to be borne by both parties. Dental hygienists should not be required to work in a manner contrary to the regulations. At best this only results in the hygienists being sceptical of the motives of the prescribing dentists and, at worst, in serious legal or disciplinary repercussions.

Based on the statutory powers conferred by the Dentists Act 1984, the legislation governing dental hygienists is laid down in the Dental Auxiliaries Regulations 1986. As with all statutory instruments, the Dental Auxiliaries Regulations 1986 make somewhat difficult reading and the General Dental Council's *Notice to enrolled dental hygienists* states these regulations in a more simple form. This notice represents the equivalent of the General Dental Council's *Notice for the guidance of dentists*.

It is expected that vocational trainees should thoroughly familiarise themselves with the GDC's *Notice for the guidance of dentists*. However, as the employment of the dental hygienist in general practice is now very commonplace, it is also important that dentists embarking on a career in general dental practice should have a working knowledge of the regulations relating to hygienists. Knowing these rules will avoid the embarrassment of being told by a dental hygienist that the prescribed treatment is not permitted within the regulations and, on a more positive note, a knowledge of the regulations may influence future practice organisation.

Enrolment of dental hygienists
All dental hygienists must enrol annually with the General Dental Council. It is important that general dental practitioners ensure that dental hygienists, acting under their direction, have maintained their enrolment. In a manner very similar to the registration of dentists, the GDC sends out a renewal notification to enrolled dental hygienists every year, indicating that the annual retention fee is payable before December 31.

Unfortunately, just as dentists may forget to reregister with the GDC, there have been occasions in recent years when dental hygienists have omitted to renew their enrolment. Such an omission could result in the hygienist being prosecuted under Section 38 of the Dentists Act 1984 for the unlawful practice of dentistry. Furthermore, paragraph 20 of the *Notice for the guidance of dentists* states:

'A dentist who employs any person to practise dentistry has a duty to satisfy himself that person is permitted by law to practise, by inspecting his practising certificate. A dentist who knowingly or through neglect of this duty enables a person to do dental work which that person is not permitted by law to do is liable to proceedings for misconduct.'

It is essential, therefore, that general dental practitioners are satisfied that all dental hygienists working under their direction are, at all times, enrolled with the GDC.

What work are dental hygienists permitted to carry out?
The Dental Auxiliaries Regulations 1986 state that a dental hygienist shall be permitted to carry out dental work (amounting to the practice of dentistry) of the following kinds:
● cleaning and polishing teeth;
● scaling teeth (that is to say, the removal of deposits, accretions and stains from those parts of the surfaces of the teeth which are exposed or which are directly beneath the free margins of the gums, including the application of medicaments appropriate thereto);
● the application to the teeth of such prophylactic materials as the Council may from time to time determine.

The statute stresses that hygienists are not permitted to carry out dental work amounting to the practice of dentistry of any other kind.

Although at first sight these regulations appear to be clearly defined, closer examination reveals that there are many grey areas which require further discussion. It is important to understand that because dentistry is subject to change and to the introduction of new techniques, statutes relating to dentistry must not be so specific as to require numerous and frequent amendments. Nevertheless these regulations provide the framework within which any interpretation of permissible treatment must be made.

In examining the forms of treatment included under the three headings listed above, reference has been made to the subject matter contained in the syllabuses for the training of dental hygienists. The General Dental Council oversees the education of dental hygienists and is responsible for publishing the *Recommendations covering courses of instruction for dental hygienists*. The recommendations indicate the minimum course of instruction which the GDC would consider necessary, in the public interest, to train dental hygienists to carry out the dental work they are permitted to do. The recommendations are reviewed at intervals and the most recent review was approved by the GDC at its meeting in May 1988. Obviously, it is fair to assume that treatments taught during dental hygiene courses would be permissible in general dental practice.

Cleaning and polishing teeth
In essence, this category is self-explanatory but confusion may arise with regard to the polishing of restorations and to what extent this may be carried out. May dental hygienists use

finishing burs and may they remove ledges? The only definitive comment that can be made is that most dental hygiene schools teach students to use finishing burs and to remove ledges with abrasive strips, finishing burs and ultrasonic scalers.

Scaling teeth
As will be shown from the full definition given in the regulations, an attempt has been made to clarify what is to be included in this category. You will note that this item is expanded to state that hygienists may remove calculus from the surfaces of teeth which are 'directly beneath the free margins of the gums'. Most schools of dental hygiene teach hygienists to carry out thorough scaling of the entire root surface within the pocket. This is in line with the view expressed by the Dental Auxiliaries Committee in September 1987: 'scaling which could be carried out through a pocket without surgical intervention was within the meaning of the regulations'. So fine scaling or root planing is permissible provided the treatment is completed through a pocket.

This section also includes 'the application of medicaments appropriate thereto'. This presumably includes substances such as chlorhexidine and desensitising agents.

Application of prophylactic materials
As this item includes 'such prophylactic materials as the Council may from time to time determine', it may be assumed that this refers to those prophylactic materials that are commonly used and covered in the training courses. The GDC has determined that dental hygienists may use any prophylactic materials including topical fluorides and fissure sealants. The permitted materials may, from time to time, be updated and it is important that hygienists and general dental practitioners keep abreast of any changes.

Items of treatment not covered by the regulations
Listed below are some of the types of treatment which hygienists may be asked to perform but are outwith those items laid down by the Dental Auxiliaries Regulations 1986.

Taking of impressions
Dental hygienists are not permitted to take impressions for any reason whatsoever.

Administering local anaesthetics
It is often argued that it would be more practical for the hygienist to administer local anaesthetics, particularly infiltrations, in those cases where patients find deep scaling uncomfortable. As the regulations stand this is not permissible and in these cases it is necessary for a dentist to give the local anaesthetic prior to scaling by the dental hygienist.

Fixing of orthodontic brackets
Techniques in fixing orthodontic brackets by chemical bonding may resemble those used in the application of fissure sealants. It is not permissible, however, for hygienists to be involved in the fixing of orthodontic brackets.

Reference to the regulations will show that hygienists can apply such prophylactic materials as the GDC shall determine. The chemicals used in preparing teeth for orthodontic brackets cannot, in this respect, be described as 'prophylactic'.

The placing of acid-etched restorations
Using a similar argument to the above, the placing of etched restorations by dental hygienists, although very similar in technique to the application of fissure sealants, is not permissible.

Temporary fillings, recementing crowns and removal of sutures
These procedures must not be carried out by dental hygienists.

The giving of oral hygiene advice
Dental practitioners reading the regulations will notice that the giving of general advice on matters relating to oral hygiene is not included among the items of dental work which may be undertaken by dental hygienists. The reason for this apparent anomaly is that the General Dental Council has been advised that the giving of general advice on matters relating to oral hygiene should no longer be regarded as part of the practice of dentistry within the meaning of the Dentists Act 1984. Such advice may readily be given by dental surgery assistants, dental health educators and other non-clinical personnel. The education of patients with regard to oral hygiene and achieving improved dental heath is a major part of the role of dental hygienists.

The dental practitioner remains responsible for any advice given to his patients but hygienists should be regarded as very able patient educators.

Must the treatment carried out by a dental hygienist be prescribed by a registered dentist?
The Dental Auxiliaries Regulations 1986 clearly state that all work performed by a dental hygienist must be carried out 'under the direction of a registered dentist and after the registered dentist has examined the patient and has indicated to the dental hygienist the course of treatment to be provided for the patient'.

This statement allows no scope for a general dental practitioner to permit a dental hygienist to formulate treatment plans. Furthermore it states categorically that the dentist must have examined the patient prior to treatment being carried out by the dental hygienist.

Any contravention of this rule could result in General Dental Council disciplinary measures being taken against the dentist and the dental hygienist.

What supervision is required for a dental hygienist working in general dental practice?
The GDC Notice to enrolled dental hygienists, based on the Dental Auxiliaries Regulations 1986, states that a hygienist working in general practice can do so 'only under the direct personal supervision of a registered dentist who is on the premises at which the hygienist is carrying out such work at the time at which it is being carried out'. This is explicit and does not leave room for misinterpretation.

The supervising dentist need not be the prescribing dentist, but it is important that a dental hygienist never carries out work without a registered dentist, who has an active responsibility for the patients concerned, on the premises. Despite the unequivocal nature of this notice, hygienists still complain of incidents where they have been expected to carry out treatment with no dentist present. No excuses would be deemed to be satisfactory for a breach of the regulations in this respect. Hygienists learn during their training that if, for any reason, a dentist is not able to be present on the premises then they must not carry out any dental work.

This ruling refers purely to supervision within general dental practice. In the Community Service or Hospital Service there is no requirement for a dentist to be present on the premises whilst the dental hygienist is working. This may be seen as illogical but the GDC notice reflecting the regulations must be interpreted as it stands.

Responsibility for work done in general practice by hygienists

The prescribing dentist has overall responsibility for the treatment carried out by the hygienist. This is supported by the requirement in the regulations for hygienists to work under the direction of a registered dentist. (The self-employed status accorded to some hygienists by the Inland Revenue is irrelevant in this respect.) In order to understand this aspect more clearly, it is helpful to examine this responsibility with regard to the GDC, National Health Service regulations and the responsibility to the patient under civil law.

General Dental Council

As both the dentist and the dental hygienist are responsible to the General Dental Council in matters of professional conduct, either party could be summoned to appear before the respective Committees within the General Dental Council and if they were found to be guilty of misconduct they could have their names removed from the register/roll. If this misconduct involved carrying out procedures not covered by the regulations, the hygienist could not use the excuse that he or she was only carrying out the instructions of the registered dentist. All hygienists should be perfectly clear in their own minds as to what treatment they are allowed to carry out. Similarly, the registered dentist would be expected to know

what forms of treatment can be carried out by a dental hygienist.

National Health Service regulations

As the contract defined by the National Health Service regulations is between the registered dentist and the Family Practitioner Committee (Health Board in Scotland and Central Services Agency in Northern Ireland), it is required that the dentist putting his name to the dental estimates form, and subsequently signing this form, takes overall responsibility for all treatment recorded. If scaling and polishing is carried out by a dental hygienist to an unsatisfactory standard, it would be the dental practitioner who would be deemed to be in breach of his terms of service. Similarly, if a prescription is laid down for a dental hygienist to carry out a prolonged course of gum treatment and the dental hygienist does not make the requisite number of appointments, the responsibility would still have to be borne by the directing dental practitioner.

Any complaint resulting in a Dental Service Committee hearing has to be made against the responsible dental practitioner. If a complaint involved treatment carried out by a dental hygienist then he/she may be called as a witness; the respondent would have to be the dentist.

Responsibility to the patient under civil law

Dental practitioners are vicariously liable for any negligence on the part of a dental hygienist committed whilst performing dental treatment under the dentist's direction. In law any claim would have to be made against the dental practitioner. Defence organisations cover dentists for any such claims.

An interesting point which may be raised in this respect is that, although there is vicarious liability, hygienists would be expected to be responsible in carrying out the treatments covered by the regulations. If it was found that the hygienist had blatantly contravened instructions, then there may be grounds for a collateral action being taken against the hygienist. As yet there is no precedent for this, but many hygienists are now personally covered by defence organisations.

8

Zoning Dental Surgeries to Reduce Cross-infection Risks

P. S. Rothwell and R. C. W. Dinsdale

Cross-infection control involves not only a knowledge of microbiological facts, but also how to apply this knowledge in a dental surgery. In the dental surgery, cross-infection control must become a way of life. The problem confronting dentists is how to create such a way of life without seriously disrupting the essential routines of clinical procedures. This is the first of two chapters addressing this problem. The first describes the development and implementation of an economical and effective zone system in conventionally-designed dental surgeries. The second will describe measures which can be taken with new surgeries where cross-infection control can be the design priority.

In the UK, cross-infection control is part of the mandate of the Health and Safety at Work Act (1974) and so implementation of adequate cross-infection control is not only an intra-professional matter. Although the Health and Safety Executive collaborate with relevant professional bodies, they have the power to act independently if necessary. Since cross-infection within a dental practice affects employees, monitoring its control falls within the remit of the Health and Safety Executive.

Measures to reduce the possibility of spreading hepatitis B during dental treatment were first reviewed in the UK in 1974 by the Dental Health Committee of the British Dental Association,[1] which was followed by the Report of the Expert Group on Hepatitis in Dentistry[2] in 1979, and in many papers since then.[3-5] More recently, advice has been concerned with AIDS.[6-10] All these papers nave focused attention on the practical steps needed to reduce the chance of a cross-infection occurring in the dental surgery, and the British Dental Association has published several general guides on procedure.[8,11,12]

Cross-infection control involves two major areas. First is knowledge of the fundamental microbiological factors which dictate the procedures needed to destroy the offending organisms. Second is the organisation of dental surgery working to implement the microbiological requirements within surgery routines. Most problems arise when a practitioner tries to follow recommendations such as those cited above because, in essence, they mean that cross-infection control must become a way of life. Difficulties are due to the interaction of several factors, particularly the following.

Nature of the work
During a clinical session in a general dental practice there is great variation in the kind of work done, requiring the use of different combinations of instruments made of different materials, some of which will not withstand the usual sterilisation procedures.

The National Health Service legacy and patient numbers
Introduction of the National Health Service in Britain resulted in huge numbers of patients seeking treatment from the small number of dentists then on the *Dentists Register*. Unfortunately, the need to treat large numbers of patients per session established the baseline for subsequent measurement of dental practice activity in the UK.

Surgery design
The majority of dental surgeries are not situated in purpose-built clinics, and their shapes and designs are not conducive to the structural modifications which, ideally, should be made.

Equipment design
Ease of cleaning, disinfection and sterilisation have not been major priorities in the design of dental equipment, and as new equipment does become available with these considerations paramount, few practices, hospitals or clinics will be able to afford to re-equip immediately or much in advance of their planned replacement programmes.

A zone system
In the current circumstances of most dental practices, a system costing little to install and run, and which causes minimal interference with normal working is worthy of consideration. This might be based upon a zone system suggested in a recent video.[13] The designation of 'clean' and 'dirty' zones is not novel and forms a central philosophy in hospital practice, but as far as we are aware, it has not been analysed in the clinical environment of dental practice.

Dentists and dental surgery assistants often touch non-sterile equipment and materials during the treatment of patients. In theory, if cross-infection is to be prevented, every item which has been 'touch-contaminated' should be disinfected after each patient. Plainly, this is not practical. An obvious alternative might be to clearly identify zones that should be cleaned when each patient leaves the surgery, the aim being to reduce the number of zones to the minimum compatible with normal working.

This chapter describes the development of a zone system in a practice setting. It does not deal with other aspects of cross-infection control, such as the provision of adequate facilities for sterilisation, sufficient handpieces, increased use of disposables and so on. These are covered expertly elsewhere.[12]

A trial

The trial was conducted in the simulated dental practice environment of the Transitional Training Unit in Sheffield University's School of Clinical Dentistry. The clinical environment of the Unit—the designs of surgeries (they are accommodated in an older family-type residence) and the equipment used—is comparable with most general practices, so from that aspect it is a valid test-bed. The critical constraint in relation to clinical work appears to be the length of the 'turn-round' time between patients (which was measured), and in this context, the fact that fewer patients are seen in comparison with general dental practice did not invalidate the results.

For one week, dental students, dental practitioners on the staff and the Unit's DSAs identified the size and situation of worktops that were required during any procedure and which equipment was regularly touched and/or moved during treatment. The handles of operating lights and bracket tables, aspirators, the dental chair and amalgam mixers were examples (fig. 1 and 2).

The number of places that were judged to be essential for staff to touch was reduced, compatible with ease of work. The worktop areas and touch-sites on equipment were tentatively marked out with coloured sticky tape or felt-tipped pen (fig. 3). For a further week, patients were treated using only these areas as the source of instruments. Some minor alterations were needed before the areas and sites were defined in the final form.

The next stage developed a treatment routine using only the designated areas, the principle being that once contact had been made with the oral cavity, everything subsequently needed during treatment should originate from only these prepared areas. Similarly, only marked areas on the equipment could be touched. This was the most difficult period as it required an accurate forecast of all instruments and materials that would be needed before treatment was started. Any omission resulted in either the dentist or DSA fetching the additional items from non-zone areas such as drawers, contaminating them in the process and wasting time with further handwashing.

When it became clear that the size and number of some instruments, such as burs, could not always be anticipated, a selection was placed adjacent to the working zone yet accessible to the seated chairside DSA using cheatles or large tweezers. These were found to be adequate for the transfer and were autoclaved afterwards. The Unit uses a tray system, but this is not essential provided that a bracket-table zone is defined for cleaning.

On completion of treatment, DSAs cleaned the designated areas of contamination, which were easily identified by the markings. For expediency, and depending on the number and variety of instruments and materials required, the DSAs often covered the worktop areas with a layer of paper towelling. They did this particularly if melted wax was likely to be present. It was easier to have amalgamators within the contamination zone, with touch-sites marked on them. Other items of mobile equipment were sited and marked, the UV lamp being an example.

Improvisation

For most equipment it was possible to devise simple measures to reduce cross-infection. For example, a small disposable polythene sleeve was fitted over the nozzles of air and water syringes and discarded after each patient. Although these items can be autoclaved, the Unit's financial position, as in general practice, precludes the purchase of a sufficient number to provide a fresh setting for each patient. The length of polythene tubing was sufficient to overcome the problem of negative-pressure aspiration of contaminated materials which is currently inherent with such syringes.

Other practical problems were encountered, and some still remain, such as handling intra-oral radiographic films. Where needed for immediate viewing, instant-type films were used and their outer cover discarded in the treatment zone without DSAs needing to rewash their hands. Where viewing could be delayed until the next appointment, the films were placed in a receptacle prior to batch processing.

The Unit's radiographic and filmprocessing equipment is in a dedicated room and it was not realistic to mark it with touch-sites. Therefore, there was no alternative to operators and DSAs washing their hands to prevent its contamination, but this should not be such a problem in practices where the radiographic apparatus is in the surgery.

Work from dental laboratories usually needs to be brought

Fig 1. Marking of operating light handle

Fig 2. Permanent marking of bracket-table handle (right)

Fig 3. The DSA's worktop marked with red tape. Note marked sites on amalgamator.

into a contamination zone. This problem was solved provisionally by spraying articulators, for instance, with an antiseptic spray prior to return to the technician.

Finally, once the system was established and running efficiently, temporary markings were replaced by permanent materials.

General dental practice

The system proved to be practical and is now in routine use. Once established, it causes little interference with normal working and, in fact, efficiency generally improves by reducing time wasted by DSAs leaving the chairside for additional materials or instruments. At first, pre-dispensing of impression materials within the worktop area was wasteful, but with experience, quantities were assessed with accuracy.

The trial was conducted in an undergraduate environment, but the system should be easier to develop in an established general practice because experienced practitioners have preferred instruments and materials, and, therefore, anticipating their needs would be much simpler. An advantage of the system is that the areas are not stereotyped. A practitioner could have as many zones as he or she desired, but the fewer and smaller, the more quickly the DSA can clean them.

It could be argued that an experienced practitioner/DSA team would not need the markings. This is true, but they reduce the time needed to teach temporary or new staff the zone system. This was proved in the Unit recently when, for another purpose, 45 complete sessions were recorded on video. Several temporary, relief DSAs were filmed, and it was clear that they adapted to the system within an hour of being introduced to it. Timed recordings showed that experienced DSAs performed the 'turn-round' cleaning between appointments within the time needed by the dentist to wash his or her hands and write notes.

Cleaning was with a chlorhexidine/isopropyl alcohol-based spray (Dispray from Stuart Pharmaceuticals Ltd, Alderley Road, Wilmslow, Cheshire) and performed between every appointment. This is beneficial because, as with brass, for example, the more often a surface is cleaned, the less effort is required. Particularly pleasing has been the displacement of

cotton wool rolls, pledgets, burs and sundry other items from bracket tables (where they constitute a communal source of supply and contamination for every patient). The problems with intra-oral radiographs could be eliminated if they were double-wrapped to offer an uncontaminated inner wrapping, which could be handled safely during transfer for processing.

During many appointments some of the designated zones will not have been touched and the temptation is to avoid cleaning them. However, the system worked best if DSAs cleaned every area and site each time, because the automatic routine this established turned out to be the major factor in reducing turn-round time between appointments. At first, the DSAs disliked the system because it seemed to impede their work, but within a short time, opinions changed, as the reduction of cross-infection risks became evident.

The five surgeries used in the trial contained identical equipment, so the marking of areas and sites was uniform except for minor variations according to the DSAs' personal preferences. However, the method has been tested in seven surgeries elsewhere that had several types of equipment, none of which precluded the introduction of the system.

Practicability

The system is a simple, practical and effective method to reduce the chances of cross-infection in dental practice. It costs nothing and the discipline imposed improves practice efficiency. We believe that if all other professional guidelines for cross-infection control are followed, such as care with sharp instruments, use of protective glasses and gloves and effective sterilisation methods,[12] practitioners introducing such a system can be confident that they are taking reasonable measures to provide a safe working environment to protect patients and staff.

In these circumstances, the treatment of the high-risk patient would require the addition of only face masks of appropriate quality and gowns (preferably disposable). Any heavily-contaminated area, such as from spillage of blood, would be treated, as appropriate, with $0 \cdot 1\%$ hypochlorite or 2% glutaraldehyde. This regimen is used in the Unit, and staff and dental students treat HIV-positive patients as a matter of routine.

Of the very many general practitioners who have visited the Unit, none has condemned the system; certainly not with the 'It would be impossible in a busy National Health Service practice' argument. Some have commented that without realising it, they had been altering their pattern of clinical activity towards the zone system, but without physically marking sites. There is scope for further progress and manufacturers of dental equipment are already responding to the challenge. For example, an operating light with removable, autoclavable handles has recently become available (Daray Lighting Ltd, Commerce Way, Stanbridge Road, Leighton Buzzard). In time, more equipment will be suitably designed, but currently practitioners will need to introduce measures that are feasible within the limitations of present environments. The system described provides effective cross-infection control in conventional surgery surroundings and, as it involves no financial outlay, it would be difficult to justify not using it or a similar method in the interests of patients and staff.

References

1 The Dental Health Committee, British Dental Association. The prevention of transmission of serum hepatitis in dentistry. *Br Dent J* 1974; **137:** 28-30.

2 Report of the Expert Group on hepatitis in dentistry. London: HMSO, 1979.

3 Scully C. Hepatitis B: an update in relation to dentistry. *Br Dent J* 1985; **159:** 321-327.

4 Samaranayake L P. Viral hepatitis: I Aetiology, epidemiology and implications, 2 Hepatitis B vaccines. *Dent Update* 1986; **13:** 353-365, 411-416.

5 MacFarlane T W, Follett E A C. Serum hepatitis: a significant risk in the dental care of the mentally handicapped. *Br Dent J* 1986; **160:** 386-388.

6 Porter S R, Scully C, Cawson R A. Acquired immune deficiency syndrome (AIDS). *Br Dent J* 1984; **157:** 387-391.

7 Richards J M. Notes on AIDS. *Br Dent J* 1985; **158:** 199-201.

8 Dinsdale R C W. *Viral hepatitis, AIDS and dental treatment.* London: British Dental Association, 1985.

9 Scully C, Cawson R A, Porter S R. Acquired immune deficiency syndrome: a review. *Br Dent J* 1986; **161:** 53-60.

10 Acquired immune deficiency syndrome AIDS booklet 3. Guidance for surgeons, anaesthetists, dentists and their teams in dealing with patients infected with HTLV III. London: Department of Health and Social Security, 1986, publication no. CMO(96)7.

11 BDA Dental Health and Science Committee Workshop. The problems of cross-infection in dentistry. Br *Br Dent J* 1986; **160:** 131-134.

12 *Guide to blood borne viruses and the control of cross infection in dentistry.* London: British Dental Association, 1987.

14 Infection control in dental practice. (Video, 16 minutes). New South Wales: Westmead Hospital, 1986.

9

Cross-infection Control as a Surgery Design Priority

L. S. Worthington, P. S. Rothwell and N. Banks

This is the second of two chapers on cross-infection control in general practice. In the first a method was described for cross-infection control which overcame the problems presented by dental surgeries of conventional design. Ideally, surgery design should allow control measures to be undertaken easily and quickly so as not to encroach on clinical time. In the course of establishing a new dental practice, an opportunity arose to make cross-infection control the design priority.

The physical features of conventionally-designed dental surgeries impede the 'turn-round' time needed for cross-infection control measures between consecutive patients' treatments. This chapter describes the development of a general dental practice which attempts to overcome the difficulties presented by such designs.

A surgery that has cross-infection control as a design priority will, to some degree, present an ambiance that conflicts with the need to offer the nervous patient a relaxed non-clinical environment; in recent years surgeries furnished to accomplish the latter have found favour. However, this reflects a time when dental treatment was often associated with discomfort or pain. The younger generation have far less experience of such traumas, which will become even less common in the future. Coincidentally, patients have become much more aware of cross-infection risks than hitherto, and this may now be more important to them than whether the surgery looks clinical or not.

Cross-infection control is a way of life, and it was realised that a radical change in the way of life in dental surgeries would dictate considerable revision of traditional working patterns. It was expected that a new surgery designed to this end would look dramatically different to patients; it was a calculated risk that their concern about cross-infection would cause them to accept the plain environment.

Cross-infection control involves operational methods as well as physical factors. Overall, the relevant issues may be grouped into four areas: design of the practice, design of the working areas, design of equipment, and operating regimen.

Current problems

Design of the practice
It is customary for not only clinical procedures to be undertaken within a surgery, but also many ancillary tasks, such as the cleaning and sterilisation of instruments, and often it serves as a store for most routinely-used materials and instruments. This creates inherent risks. For instance, clean and dirty instruments may be in close proximity and the collection and cleaning of dirty instruments undertaken between the change-over of patients, when a DSA might be involved with other duties, such as instructing the receptionist about the next appointment. More importantly, if appointments are running late, the DSA may be under pressure to speed up collection of dirty instruments, yet most 'stick' injuries from sharps occur during this activity. Thus, the procedure which should be performed under conditions of greatest calm is frequently undertaken in periods of greatest pressure.

Design of the surgery area
Surgery design hinders cross-infection control mostly because it aims to fulfil the multipurpose roles mentioned above. Even the most streamlined units, cupboards and other non-clinical furnishings are difficult and time-consuming to clean, and moving materials and instruments stored in surgeries absorbs considerable clinical time. Frequently, the equipment and controls used by the dentist during clinical work are situated on the opposite side of the dental chair from those required by the DSA, and this duplicates the areas which need cleaning between operating periods.

Other problems in the surgery involve the types of materials used in its construction, and in the control of the environmental air.

Design of dental equipment
The design of equipment often precludes ease of decontamination, but manufacturers are recognising this deficiency and designing accordingly. In addition, the evacuation of contaminated air, water and other waste materials is subject to strict regulations, and the supply of uncontaminated air and water to equipment presents problems.

Operating regimen
The zone system described in Chaper 8 outlined a complete revision of operating methods, whereby the instruments and materials that would be needed were anticipated rather than selected on an ad hoc retrieval basis which necessitates repeated handwashings. Cross-infection control demands critical assessment of all operational routines.

Solutions attempted in the new practice
From consideration of the problems cited, the design of the practice was based upon a logical extension of the zone system. That is by removing zones of potential contamination from the surgery.

Design of the practice
It was clear that the major requirement was to remove from the surgery anything that is not required clinically, thereby reducing the sites to be cleaned. The most obvious items were the storage areas, and as they include almost every cupboard, unit and drawer, considerable surgery space would be saved

which could be put to other purposes. This would result in surgeries that were comparatively small, but sufficient for clinical needs and movement of patients and staff. Although the surgery furnishings would be sparse, this could be countered by the creation of an attractive dental health education section in the saved surgery space.

The central feature would be a sterilisation/cleaning room to serve each surgery via hatches, thus removing instrument cleaning and sterilisation procedures from the surgeries.

The plan which evolved is shown (fig. 1). Although two floors were available, the plan outlines only the ground floor, where the relevant clinical areas were sited.

The central sterilisation/cleaning room is fully equipped with autoclave, ultrasonic cleaner, pre-sterilisation handpiece

cleaner, a safe needle-remover and so on. It also offers a rapidly accessible site for the ready-to-use kit for medical emergencies. As this is a non-surgery area, the basic furnishings are low-cost kitchen units. The photograph (fig. 2) shows the working end of the room and the storage racks for the instrument trays used in the system. As a temporary measure, prior to planned expansion to the second floor, the radiographic orthopantomograph is installed behind a lead-lined screen in that room but distant from the sterilisation and cleaning section.

The health education area (fig. 3) is fitted with three low-level sinks and a long wall mirror and divided from the reception/ waiting area by a large mobile screen. This enables parents to hear their children's dental hygiene advice and,

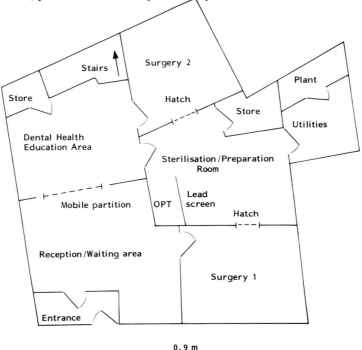

0.9 m

Fig. 1 Outline plan of the practice.   ├───────┤

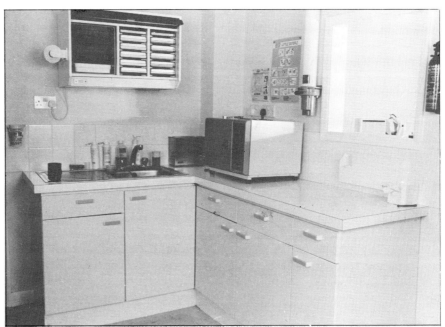

Fig. 2 The sterilisation/cleaning room showing a hatch (right) to a surgery and a storage rack for the tray system.

Fig. 3 The dental health education area. The mirror reflects dental health posters on the back of the mobile screen.

indirectly, absorb it themselves.

A central plant room is installed with equipment which will service future expansion.

Design of the surgeries

This was considered in three categories: overall plan and construction; environmental air; and the introduction of air and water, evacuation of waste materials, and the selection of dental equipment.

Overall plan and construction

It was possible to confine all clinically-required equipment to the area behind the dental chair. Thus, except for the wall-mounted sinks on the side walls behind the chair area, the wall behind the chair is the only one in the surgery that is not completely plain. This offers ideal conditions for rapid and effective cleaning. On the wall behind the chair are mounted the light-curing lamp (to preclude constant contamination of the whole unit), the radiograph viewer, the dispensers for paper towels and handwash solutions and the clinical-wipe tissues supply.

For the same cost as conventional flooring, material was layed which possesses intrinsic bactericidal properties (Krommenie Marmoleum—Forbo-Nairn Ltd, Leet Court, 14 King Street, Watford, Hertfordshire).

Environmental air

It is desirable to regularly evacuate environmental air because of the pathogens it contains, but at the same time to maintain optimum ambient temperature and humidity. In preference to central heating, independent wall-mounted electrical units were installed, which use a reverse heat pump system (Toshiba RAS22GKH—Thermotech, Unit 38, Works Road, Hollingwood, Chesterfield). This is economic yet fulfils the above criteria. Furthermore, the air in the surgeries is maintained at a slightly positive pressure and thus expelled from the surgery.

The introduction of air and water, evacuation of waste materials and the selection of dental equipment

These are interrelated problems. Most were solved by a specific trolley unit (Adec Excellence Cart—Adec UK Ltd, 75 Gravelly Park Industrial Estate, Birmingham). Although relatively expensive, it serves both the dentist and the DSA, halving the cleaning operation compared with cleaning a dual system, as mentioned previously. The single unit also reduced electrical and plumbing installation costs and it contains several storage sections which can be used to house the small bottles of lining materials, cotton wool rolls and so on. The water introduced for the handpieces and three-in-one syringes is sterile, and their air supplies are introduced via a breathing air filter. An overview of the operating area of a surgery is shown (fig. 4).

The operating lamp (not shown) has detachable, autoclavable handles, and a smooth clear screen, which is hinged so that it can be folded in front of the lamp to make cleaning debris away easier (Daray Lighting Ltd, 7 Commercial Way, Stanbridge Road, Leighton Buzzard, Bedfordshire).

Evacuation of waste is by large-volume wet-line central suction (Den-Tal-Ez Dental Products GB Ltd, Cleveland Way, Hemel Hempstead, Hertfordshire). Waste is passed to the plant room and then into external sewers. It must go there rather than to routine water drains. There is a filtered exhaust vent externally which, again, is mandatory.

All waste from the practice is considered to be clinical waste and disposed of according to local authority regulations which conform to those of the Health and Safety Executive.

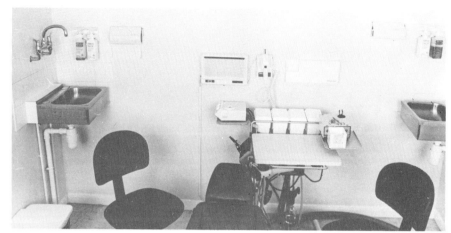

Fig. 4 Overview of the operating area of a surgery.

Operating regimen

Obviously, the central sterilisation room requires staffing. When the practice was newly opened and operating only one surgery, this was not difficult because the patient throughput allowed a DSA to work both this room and the reception area. However, when two-surgery running came into operation, it became clear that the cost of staffing the sterilisation room was far outweighed by the financial and operational advantages to the practice.

● The surgery DSAs can devote undivided attention to chairside duties, and this increases clinical efficiency.

● As the surgery DSAs do not clean and sterilise instruments, they are not at risk due to haste when surgeries are busy. The timing of these procedures can be 'buffered' in the sterilisation room and, thus, always undertaken at an optimal pace. This is a very important benefit.

● The DSA working in the sterilisation room does not restrict activities to sterilisation procedures; she operates the OPT machine, processes radiographs and mixes impression materials which she passes in ready-loaded trays to the surgeries. The completed impressions are then returned from the surgeries to be washed and decontaminated.

● Surgery cleaning between consecutive patients' appointments is approximately 30 seconds, thereby solving the major problem in terms of the financial implications of cross-infection control, namely, 'turn-round' time.

● Efficiency is increased due to the simplicity of maintaining the equipment systems, and the reduction in overall routine cleaning procedures.

● The sterilisation/cleaning room provides an excellent training station for trainee DSAs. Introduction to instrumentation, the tray system and sterilisation procedures, combined with surgery-assisting during the more straightforward clinical procedures, offers a first-class basic training to new employees. Yet, experienced DSAs working this room on a rota enjoy the variety of this occasional change of duty.

Costs

Many costs of cross-infection control, such as those of disposables, are always incurred. In the present system, there are two areas where costs differ from the conventional arrangement; first, the capital costs of establishing the practice and second, the running costs.

Table I shows the comparative costs of a new system and a conventional layout, as prepared by a commercial dental company (Healthco (UK) Ltd, 7-8 Gauntley Court, Redford Road, Nottingham). The costs were based upon identical equipment and instruments in each situation. These are variable, of course, but a standard costing permits a true comparison between the conventional and current systems. Figures quoted are inclusive of VAT. The overall percentage differences will probably apply to any price changes.

For a single-surgery installation the total cost is approximately 5% more than that for a conventional system. However, with two surgeries, it is roughly 10% less and for three surgeries, some 20% less than the conventional costs. The major savings with the increase in numbers of surgeries are due to the reduction in duplication of instruments as a result of a central sterilisation-room supply, and the reduction in surgery cabinetry.

On a single-surgery basis, the running expenses are increased by the staffing of the sterilisation room, but this is less than the expense incurred by loss of clinical time to effect adequate cross-infection control procedures in a conventional surgery. With two or more surgeries, the costs of staffing the sterilisation room are reduced according to the number of surgeries serviced.

Approximately 3 weeks were needed to adapt to the operational routine. However, any major change to combat cross-infection risks will involve a period of transition.

Patient response

There is no evidence that the rather bare, uncluttered surgery surroundings have dissuaded patients from attending, and the impression we have gained is that attention to the furnishings of the reception/waiting room, together with the novelty of the health education area, have compensated for the surgery surroundings. Many patients have remarked upon the very obvious measures taken to control cross-infection.

The practice opened as a squat, with one surgery, in October 1987. Its situation was selected after consultation with the district dental officer to locate a city area where there was a dental need. By December 1987 appointments were fully booked for an acceptable period, and in January 1988 a second surgery was complete. In that month it was appointed by the North Trent dental vocational training scheme as a training practice and a trainee was placed. Although the considerations for setting up the practice were not specifically directed towards cross-infection control, its steady growth has

Table I Comparison of costs of dental equipment in 'conventional' and 'cross-infection control' surgeries. Costs (£) include VAT.

	'Conventional' (A)	'Present' (B)	(B–A)
One surgery*	35 528	37 432	+1904
Two surgeries	55 290	49 772	−5518
Three surgeries	75 054	60 432	−14 622

* Initial costs include one orthopantomograph and one compressor. All costs include all dental equipment, instruments and one radiographic unit per surgery.

dispelled any doubts that patients would be daunted by surgeries which looked clinical rather than inviting.

If dental practitioners are not to be handicapped by trying to practise cross-infection control within difficult environments, future plans will demand reappraisal of basic concepts. This practice does not illustrate the only solution; it is described in the hope that practitioners will find the features and experiences outlined helpful in any future planning.

10

Viruses in the Aetiology of Cancer

C. Scully

Viruses are achieving greater importance, as the role of known viruses in diseases is increasingly appreciated and new viruses are recognised. There are clear causal relationships of viruses with some animal malignancies and there has been speculation as to the role of viruses in human neoplasia. The dental surgeon should be aware of the current concepts in this field, particularly in relation to oral neoplasms. This chapter briefly reviews the state of the art at a basic but up-to-date level. It is not intended to be fully comprehensive and therefore a further reading list of recent papers is appended for those who might wish to delve deeper into the field.

Viruses contain nucleic acids but, in contrast to other microorganisms, they contain virtually only ribonucleic acid (RNA) or deoxyribonucleic acid (DNA). Viruses are ubiquitous and generally cause subclinical (asymptomatic) infections, or disease of little consequence to most individuals.

Some viruses have been incriminated, however, in the aetiology of human neoplasms. This raises the obvious but fundamental question that if viral infections are so common, how can one distinguish whether a virus associated with a neoplasm is causing the lesion or simply infecting the tissue with no malignant consequence? Another obvious question is why should the same virus produce an inconsequential result in one individual but, in another, an often lethal tumour? If the virus has not changed, clearly either the hosts differ in their response to the infection or cofactors may be acting in the one instance to produce malignancy.

Evidence required to incriminate a virus as a cause of a neoplasm (Table I)

The various parameters (Koch's postulates) required to implicate a bacterium as the causative agent of a disease are well known to dental surgeons, but the only definitive proof that a virus causes a neoplasm is if immunisation or other prophylaxis against the infection regularly protects against the tumour development. Other evidence is indirect.

Only very rarely are intact viral particles detectable in tumours, but this does not mean that the virus has not been, or is not, present. The virus may have previously infected the cell and caused a mutation (known as the 'hit and run mechanism'), or it may have 'hit and stayed' but only be detectable at a molecular level. Its presence may be detectable because viral proteins (antigens) can be found as 'footprints' of the virus, but some viruses are present in a cell and not

actually transcribing the messenger RNA (mRNA) needed to code for protein production. It can be seen, therefore, that neither the absence of intact virus, nor the absence of viral proteins, prove that there is, or has been, no virus present.

The presence of some viruses associated with neoplasia can be established because viral nucleic acid is present in the cell nucleus: this may be slotted into a host chromosome (that is, 'integrated') or may be floating free ('episomal'). It is more obvious how integrated viral DNA could affect host genes so as to cause neoplasia than how episomal viral DNA could. The presence of viral DNA in a cell, therefore, is suggestive of, but is alone not proof of oncogenicity.

Animal and cell culture studies may strongly suggest that the virus in question is oncogenic in those systems, but this cannot necessarily be extrapolated to man. For example, some papilloma viruses cause oral or oesophageal cancer in cattle but are not known to harm man. Herpes simplex virus can transform cells in culture, but there is no proof that it is oncogenic to man (see below).

Epidemiology of the human tumour may help identify the causal agent, and classic examples of this are African Burkitt's lymphoma, which is associated with Epstein-Barr virus (EBV) and coincides strongly with endemic malaria areas; and primary liver cancer which has a very strong association with hepatitis B virus (HBV) infection and other cofactors (see below). Clusters of patients developing the tumour may also provide some evidence.

Examination of serum antibody responses to the virus in question may provide further circumstantial evidence of an association. For example, serum levels of IgA antibody against certain antigens of Epstein-Barr virus (EBV) relate extremely closely to the prognosis in nasopharyngeal carcinoma. Nevertheless, it is important, although difficult, to exclude mere chance associations such as geographic and socioeconomic coincidences.

Correlation is not causation. For example, there is a vast literature on sero-epidemiology purporting to show that herpes simplex virus (HSV) genital infection predisposes to cervical cancer and yet the same women who have been infected with HSV are likely to have other sexually transmitted diseases that may be incriminated, such as human papillomavirus (HPV). Furthermore, they often smoke and may have been more exposed to putative carcinogens in, for example, semen. In such studies it is crucial, therefore, that all possible factors are controlled and controls must clearly be of the same age, sex, socioeconomic status and so on.

Table I Evidence towards incriminating a virus in oncogenesis

1 Is the virus oncogenic to animals?
2 Does the virus cause cell transformation?
3 Is infection with the virus usually followed by appearance of tumours?
4 Is there sero-epidemiological evidence of an association with the tumour?
5 Are there 'clusters' of tumours?
6 Are viral antigens or nucleic acid present in the tumour?
7 Does immunosuppression predispose to the tumour?
8 Does immunisation against the virus prevent the tumour?

Evidence suggesting why viruses are not consistently oncogenic

The usually rare but often important 'experiments of nature', such as patients with genetic defects who appear predisposed to certain neoplasms, may help provide evidence in this respect. For example, patients with a rare genetic skin disorder, epidermodysplasia verruciformis, are predisposed to skin carcinoma caused by selected HPV types that appear to be innocuous to the general population.

In other persons, interference with immune responses— such as in patients on corticosteroids, or in AIDS— predisposes them to tumours which may be virally induced, such as by EBV. Further examples are the genetic factors and possible environmental factors (for example, dietary carcinogens such as nitrosamines) that predispose certain groups infected with EBV to develop nasopharyngeal carcinoma (for example, the Chinese) (fig.1).

Finally, there can be a tremendous delay between infection with the virus and the appearance of a tumour (Table II), and this sometimes makes it difficult to appreciate that there is an association.

Viruses and non-oral neoplasms (Table II)

Epstein-Barr virus (EBV)

EBV is a herpes virus. The herpes virus group are DNA viruses that are characterised by the fact that, after primary infection, the viruses remain latent in the body (and can be reactivated).

Primary infection with EBV, if it causes symptoms, presents as infectious mononucleosis (glandular fever). Thereafter, the virus is latent in pharyngeal epithelial cells and shed in saliva. Chronic symptomatic EBV infection is also a recognised entity.

Infection with EBV early in life, in persons with an appropriate genetic make-up and in the presence of other cofactors, may lead to malignancy—Burkitt's lymphoma, lethal midline granuloma, or nasopharyngeal carcinoma. The cofactors in African Burkitt's lymphoma appear to be holoendemic malaria, which presumably acts by depressing the T lymphocyte immunity needed to control EBV, while also causing B lymphocytes to proliferate. In immunocompromised patients in other geographical regions, EBV can lead to malignant lymphomas.

At the molecular level, there are chromosomal changes such that a cancer gene (an 'oncogene' called c-myc) is translocated and comes to lie in the middle of one of the genes that control immunoglobulin production (on chromosomes 2, 14 or 22) and thereby in some way activates growth of B lymphocytes. The combined effect of these various insults on the B lymphocytes seems to be to cause their malignant transformation and Burkitt's lymphoma.

The cofactors in nasopharyngeal carcinoma appear to be chemicals—probably dietary nitrosamines, possibly diterpene esters from plants (Euphorbaceae)—and possibly habits such as use of nasal sprays. The molecular basis for nasopharyngeal carcinoma is unclear.

Hepatitis B virus (HBV)

HBV is a DNA virus that, if it causes symptoms causes hepatitis. Most infected patients clear the virus with little consequence, but a few become carriers. A number of case-controlled studies in Greece, Senegal, USA, Japan and Taiwan have shown a close association between primary liver cancer and HBV infection. Cofactors such as fungal toxins (aflatoxin) from contaminated peanuts, corn, rice, beans or alcohol, cigarette smoking, immuno-suppression, and alcohol itself, markedly increase the risk. Indeed, in Japan, smoking is more closely associated with liver cancer than with lung cancer.

It remains to be seen whether the incidence of liver cancer will be as significantly reduced by vaccination programmes against HBV such as those in China as expected, but there is every reason to be optimistic.

Human papillomaviruses (HPV)

HPV are DNA viruses that typically cause benign epithelial neoplasms such as various warts. Over 50 HPV types are now recognised.

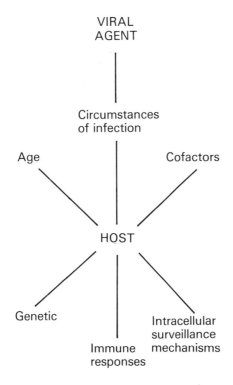

Fig. 1 Factors influencing outcome of virus infections.

Table II Oncogenic human viruses

Viruses	Common manifestations of infection	Neoplasms associated with viruses	Latency to development of neoplasm (years)
Epstein-Barr virus	Infectious mononucleosis	Burkitt's lymphoma	3–12
		Nasopharyngeal carcinoma	30–50
Hepatitis B virus	Hepatitis	Primary liver cancer	30–50
Human papillomaviruses			
Types 5,8,14,17	Warts	Skin cancer in epidermodysplasia verruciformis	3–12
Types 16,18,31,33,35	Warts	Anogenital neoplasms	20–50

The oncogenic potential of certain types such as HPV-5 and HPV-8 in genetically predisposed patients with epidermodysplasia verruciformis has been mentioned above. Cofactors such as ultraviolet light, irradiation or immunosuppression increase the frequency of malignant transformation. Another HPV type, HPV-41, however, appears to be associated with skin cancer in apparently normal individuals. Anogenital neoplasms appear especially related to infection with HPV types 16, 18, 31, 33 or 35. Cofactors, possibly HSV or smoking, may be involved.

The molecular basis of oncogenesis appears to be a disruption or rearrangement within the HPV gene at a specific site that should normally regulate two of the HPV genes required for tumour production (early genes E6 and E7).

Human retroviruses

The human retroviruses are RNA viruses that transcribe from RNA to produce DNA, that is the reverse of normal transcription. They contain an enzyme which is therefore called 'reverse transcriptase'. A number of these viruses are known, but some are still 'searching' for a disease. It is unclear which, if any, disease they produce. The most important retroviruses known at present are the human T lymphotrophic retroviruses (Table III), which include HTLV-1 (associated with adult T cell leukaemia); HTLV-2 (associated with T cell hairy cell leukaemia) and the various human immunodeficiency viruses (HIV).

It is interesting, in view of the concern about HIV, to note that intravenous drug abusers are often HIV positive, and about 6% of intravenous drug abusers in London are HTLV-1 positive!

Other viruses

The range of other potentially oncogenic viruses that are appearing on the horizon are shown in Table IV. Solid evidence of oncogenicity is at present lacking.

Viruses and oral carcinoma

Three viruses have been implicated in oral cancer, namely HSV, HPV and adenoviruses.

HSV

Anecdotal clinical evidence suggested that some lip carcinoma followed recurrent herpes labialis, but this association is highly dubious. Sero-epidemiology has shown raised titres of antibodies to HSV in oral cancer patients but related mainly to their smoking habits. Neither HSV antigens nor HSV-DNA have been demonstrated in the tumours and, if HSV is involved, it would seem to be by a 'hit and run'

mechanism. Certainly, HSV has oncogenic potential *in vitro* comparable to some potent chemical carcinogens.

HPV

HPV antigens and HPV DNA have been found in oral carcinoma and in premalignant lesions and, as discussed above, certain HPV types are oncogenic at other sites.

HPV are, however, widespread in the normal oral mucosa and in benign lesions and, therefore, if they are associated with carcinoma, it would seem they must be acting in concert with cofactors such as smoking, alcohol or other agents, or that the aetiology of individual tumours can vary. An unusual HPV type related to HPV-16 is found in oral squamous carcinoma but, in verrucous carcinoma, HPV-2 is the main agent. HPV are implicated in several benign oral warty lesions.

Adenoviruses

These have now been discounted in the aetiology of oral carcinoma.

Summary

There is strong evidence for associations of EBV with some types of Burkitt's lymphoma and nasopharyngeal carcinoma and of HBV with liver carcinoma. There may be associations of HPV with anogenital or oral carcinoma but these remain to be proven. A wide range of viruses have a potential for oncogenicity and further work is clearly required in this area.

Further reading

Crawford L. Criteria for establishing that a virus is oncogenic. In *Papillomaviruses*. pp 104-116. Ciba Foundation symposium. Chichester: Wiley, 1983.

De The G. Prevention of virus-associated human malignancies: an epidemiological view. In Maskens A P (ed). *Concepts and theories in carcinogenesis*. pp 311-321. Amsterdam: Elsevier Science Publishers, 1987.

De Villiers E M, Neumann C, Le J Y, Weidauer H, zur Hausen H. Infection of the oral mucosa with defined types of human papillomaviruses. *Med Microbiol Immunol* 1986; **174:** 287-294.

Epstein M A, Achong B G (eds). *The Epstein-Barr virus: recent advances*. London: William Heinemann Medical Books, 1986.

Haase A T. Pathogenesis of lentivirus infections. *Nature* 1986; **322:** 130-136.

Harabuchi Y, Yamanaka N, Kataura A *et al*. Epstein-Barr virus in nasal T cell lymphomas in patients with lethal midline granuloma. *Lancet* 1990; **335:** 128–130.

Klein G (ed). *Advances m viral oncology*. Vol 7. New York: Raven Press, 1987.

Anonymous. Genital warts, human papillomaviruses and cervical cancer (Leading article). *Lancet* 1985; **2:** 1045.

Table III Human T lymphotrophic retroviruses

	Viruses	Associated diseases
Oncoviruses	HTLV-1	Adult T cell leukaemia Tropical spastic paraperesis
	HTLV-2	T cell hairy cell leukaemia
Lentiviruses	HIV-1*	AIDS
	HIV-2	AIDS

*Formerly HTLV-3

Table IV Other viruses of possible oncogenic potential

Viruses		Comments
Adenoviruses 12,18,31		Oncogenic in rodents but not known to be in man
Polyomaviruses	BK	Gliomas Ependymomas Insulinomas
	JC	Leiomas in monkeys Leiomas in man?
	Lymphotrophic polyomavirus (LPV)	Oncogenic to hamster cells but not known to be oncogenic to man
Human B lymphotrophic virus (HBLV) (also known as lymphotrophic herpes virus LHV)		AIDS? Lymphoproliferative disorders? Sarcoidosis?

Maitland N J, Cox M F, Lynas C, Prime S S, Meanwell C A, Scully C. Detection of human papillomavirus DNA in biopsies of human oral tissue. *Br J Cancer* 1987; **56:** 245-50.

Nishioka K. Hepatitis B virus and hepatocellular carcinoma: postulates for an aetiological relationship. *Adv Viral Oncol* 1985; **5:** 173-199.

Salahuddin S Z, Ablashi D V, Markham P D *et al.* Isolation of a new virus HBLV, in patients with lymphoproliferative disorder. *Science* 1986; **234:** 596-601 .

Scully C, Prime S S, Maitland N J. Papillomaviruses: their possible role in oral disease. *Oral Surg* 1985; **60:** 166–174.

Scully C, Cox M, Maitland N J, Prime S S. Papillomaviruses: the current status in relation to oral disease. *Oral Surg;* **65:** 526–532.

Scully C, Samaranayake L P. *Viral infections and dentistry.* Cambridge: Cambridge University Press (in press).

Trichopoulos D, Day N E, Kaklamani E *et al.* Hepatitis B virus, tobacco smoking and ethanol consumption in the etiology of hepatocellular carcinoma. *Int J Cancer* 1987; **39:** 45-49.

Weiss R A. Molecular and cellular aspects of retrovirus pathogenesis. In *Molecular basis of virus diseases.* pp 167-192. Cambridge: Cambridge University Press, 1987.

11

Oral Surgery: Staying Healthy

R. L. Caplin

As one of the caring professions, dentistry has always concerned itself with the welfare of its patients, tending to their aches and pains, their needs and demands, and treating and preventing those diseases that occur in and around the mouth; we offer our skills and expertise to improve the quality of life of others. Unfortunately, nowadays the provision of such a service to our patients carries with it risks that were unknown 25 years ago. Once, when antibiotics were effective against syphilis and tuberculosis, dentistry carried little risk to the health of the practitioner and his staff. With the emergence of diseases that pose severe threats to both dentist and staff, our concept of a caring profession must now be broadened to include not only caring for the patient, but also caring for the health and welfare of the dentist who provides the treatment and for the staff who support him in this provision.

There is an inter-relationship between dentist, patient, staff and equipment and a 'caring' professional must look at all of these aspects to preserve and maintain the health of all who pass through the surgery. Patients attending the surgery will be in varying states of health, both physical and mental, and it is of vital importance that the practitioner recognises and understands the impact of his treatment on the well-being of the patient. The intention must be that the patient leaves the surgery in the same state of general health as when he entered, and is not compromised either by the treatment given by the dentist or by the transmission of infectious diseases. Consequently, a thorough medical history must be obtained from all new patients attending the surgery, even if it is for a 'non-invasive' procedure, such as refixing a loose crown. The medical history of recall patients should be updated at the beginning of each new course of treatment.

There are various ways of obtaining and recording a medical history. Perhaps the most effective is to give the patient a prepared form to fill in while they are waiting to be seen. This will give them time to remember their previous medical and surgical experiences and will also prompt them with relevant questions. The completed questionnaire is then kept with the patient's records, having been checked through by the dentist and any suspect points further expanded by direct questioning. Alternatively, the dentist can question the patient directly and note the answers on the record card. This does have the advantage of helping to establish a relationship and of enabling the dentist to assess and observe the patient, but it does tend to be a rushed affair and, in the short time available, the patients often forget their previous experiences. For example, women with children will often say they have not been into hospital, forgetting the times when they gave birth to their children, or assuming that it is not relevant. Therefore, the worry must always be that items which are of great significance will not be disclosed by the patient, because they are considered by them to be unimportant.

Having taken a detailed medical history, it is essential to make a complete entry in the notes; 'nil relevant' is uninformative, giving no indication of the questions asked or the answers given. Depending on the history given, adjustments to the treatment plan may have to be made, to

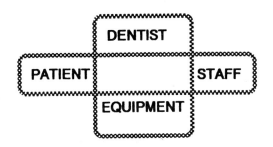

CONFIDENTIAL HEALTH QUESTIONNAIRE

To obtain the best and safest treatment, your dentist needs to know of any problems which may affect your treatment. Please only tick the box if the answer is YES.

Date .... Sex .... Age ....

Surname .... First Name(s) ....

Occupation.................................... Address....................................

Your Doctor's Name & Address ....

ARE YOU:
1. Attending or receiving treatment from a doctor? ....
2. Taking or using any medicine, pills, tablets, inhalers, ointments, injection or any other drug?.......
3. Allergic to or ever had any bad reaction to any medicines — particularly antibiotics, foods or other substances?....................................

HAVE YOU :
1. Had rheumatic fever or chorea (St Vitus' Dance)? ....
2. Had jaundice, liver, kidney disease or hepatitis? ....
3. Ever been told you have a heart murmur or heart problem? ....
4. Had any blood pressure problems? ....
5. Had any blood tests, inoculations etc? ....
6. Had a bad reaction to a general or local anaesthetic? ....
7. Had a joint replacement? ....

DO YOU:
1. Have a pacemaker or have you had any form of heart surgery? ....
2. Suffer from hay fever, eczema, asthma or any other allergy? ....
3. Have chest problems? ....
4. Have fainting attacks, giddiness, blackouts or epilepsy? ....
5. Have diabetes or does anyone in your family? ....
6. Bruise easily or following extraction, surgery or injury have you or your family bled so as to cause you to be worried? ....
7. Carry a Warning Card?....................................
8. Think you may be pregnant? ....

Are there any other aspects concerning your health that you think the dentist should know about? ....

Completed by: Self/Parent/Guardian Signature.................................... Date ....

Have there been any changes in your health, medicines, injections or tablets since your last course of treatment? ....

Date.................... Date.................... Date.................... Date....................

© British Dental Association 1987

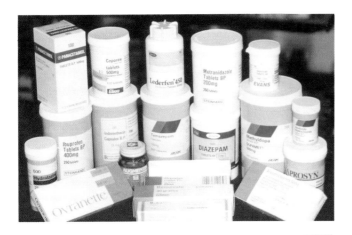

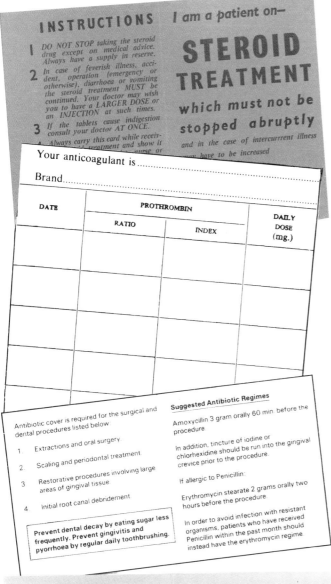

ensure that the general health of the patient is disturbed to the minimum. Many patients will be under the care of a physician or surgeon and may be taking medication either in the short term or the long term. Long-term conditions that are commonly encountered are hypertension, cardiac complaints, joint problems, skin conditions, epilepsy and depression. Short-term conditions being treated are generally infections.

If further information about a patient's medication is required, it is always possible to contact the patient's general medical practitioner, as well as looking in authoritive publications such as the *British national formulary, MIMS,* or the *Drugs and therapeutic compendium.* Here, the side effects of the drug and any possible drug interactions will be listed, and it is important to remember that the local anaesthetics that we use, together with their additives, such as adrenaline or noradrenaline, are drugs with the potential to interact with the patient's current medication to produce undesirable, or even alarming, reactions.

Some patients carry cards, informing the dentist of the condition they are suffering from, their current drug regimen and precautions that the dentist must take when treating them. Steroid and anticoagulant therapy fall into this category, and it is advisable to contact the person supervising this aspect of the patient's treatment (either the general medical practitioner, the consultant or the haematology department concerned) to get advice as to how best to perform your dentistry within the framework of their current medication.

Some patients will have cards advising of the need for prophylactic antibiotic cover (for example, prosthetic valve replacements, hip replacements), and the practitioner will then be alerted to the need to protect the patient with the currently recommended regimens. If you are uncertain or more information is necessary, do not be afraid to contact the person supervising the patient; it is better to ask nine times than to go wrong once.

Apart from a health risk to the patient from treatment, there is also a health risk to the dentist from the patient; again, the medical history should alert the dentist to any potential or real problems in this respect. As dentists, we work in an environment which at the least is dangerous and at worst could be lethal. We are exposed to a number of potential illnesses, which may incapacitate for days, weeks, years or even for life. A few questions and adequate precautions will save much pain, heartache and misery.

Hepatitis B and HIV infections have emerged over the past quarter of a century as the most harmful of the infections to which we are exposed, and it is clear that infection with either of these can have disastrous consequences. With the contraction of hepatitis B, an absence from the surgery of 6 months or more is likely, together with loss of income and goodwill. In some cases, it has been necessary for the dentist to sell the practice in order to avoid financial ruin. Even in the absence of the overt disease, an HIV positive or hepatitis B carrier dentist would face a moral dilemma with regard to treating his patients. Hepatitis B, as a transmissible disease, is in fact much more of a risk than HIV, as is well expressed in the following way: if a teaspoonful of HIV virus was poured into two pints of water and one millilitre of this was injected into an individual, the person would contract AIDS. If a teaspoonful of hepatitis B virus was poured into a 25 000 gallon swimming pool and one millilitre of this was injected into an individual, that person would contract hepatitis B.

It is sometimes difficult to know from a medical history whether a patient's history of jaundice was in fact hepatitis A or B, and rather than refer every suspect patient to the local pathology laboratory for investigations, which can be expensive and time-consuming, further questioning may give a clearer idea as to which type of hepatitis it might have been. Initially, of course, contacting the patient's general medical practitioner should reveal much relevant information.

Typically, jaundice between the age of 15 and 45 years, of several weeks, duration and followed by several weeks or months of weakness and general ill health, the jaundice occurring as an isolated case as opposed to one of several cases and involving, perhaps, hospitalisation, are pointers to hepatitis B rather than hepatitis A. Known carriers of transmissible diseases should be offered treatment in the dental surgery and this is best performed at the end of the session.

Coughs, colds and influenza are other infections easily passed on to the dentist; the wearing of a mask may help to prevent their transmission from patient to dentist. If worn, masks should be changed at frequent intervals. Specific infections such as syphilis and TB also present a risk, albeit rare, and any ulcer should be treated carefully and with gloved hands. 'Cold sores' (herpetic lesions) are frequently found around the mouth and lips of patients, and the open sore is quite likely to pass herpes virus on to the fingers of the dentist. This can produce a painful and debilitating herpetic whitlow, which will almost certainly lead to an absence from work, as they are as painful as they look!

Aerosols produced by the air turbine, the three-in-one syringe and the ultrasonic scaler will contain organisms and it is prudent to protect against their potentially harmful effects by wearing glasses and a mask. Simple precautions are therefore needed to prevent illness, or even death, while the daily routine of dentistry is carried out.

In the inter-relationship between dentist, patient and staff, it is clear that organisms can be transferred not only directly but via deposition and pick-up from the instruments, utensils and surfaces that the individuals may come into contact with. Protection against this transfer may be considered as primary, in the prevention of direct exchange of organisms during contact with body fluids, and secondary in the negation of the harmful effects of organisms, either by boosting the body's defences by vaccination or rendering the organisms ineffective by sterilisation.

Direct contact with body fluids such as blood, saliva, tears, mucus and pus, any of which may contain potentially pathogenic organisms, must be avoided. The usual portal of entry is via a break in the skin of the hands, or perhaps via a break in the integrity of the mucous membranes of the eyes, the respiratory tract or the gastrointestinal tract. Skin protection is provided by wearing protective gloves and today there is such a wide variety of sizes, quality and price, that every practitioner will be able to find gloves to suit. These may be worn for as long as they are reasonably comfortable and intact, and can be washed between patients with soap (preferably liquid). Some makes will in fact be usable for several sessions before deteriorating beyond reasonable comfort. Sweating of the gloved hand is a problem which worsens as the glove is worn for longer periods; coating the

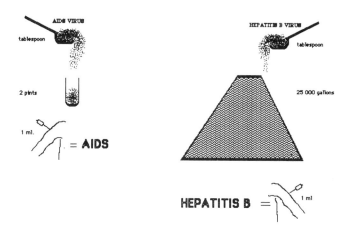

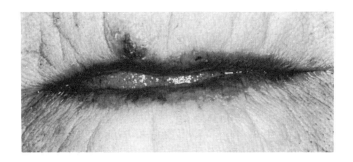

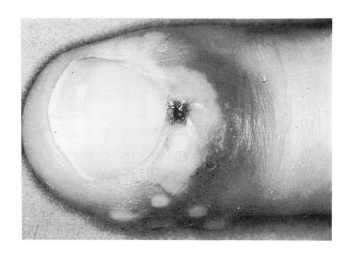

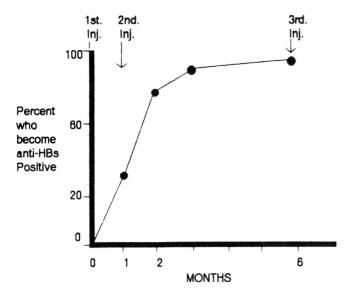

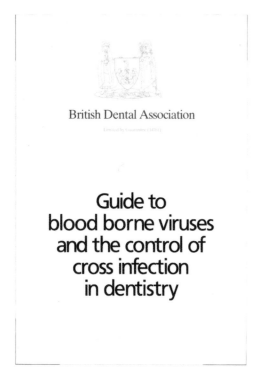

British Dental Association

Guide to blood borne viruses and the control of cross infection in dentistry

hand with ordinary talc or starch powder prior to donning the gloves will help, as there is usually insufficient powder supplied with the gloves themselves.

Protective glasses should be worn by dentist, nurse and patient, to protect the eyes, and masks should be worn by the dentist and the nurse to prevent the inhalation of particles during the production of aerosols, either from the turbine or the air syringe. It is sensible to change masks frequently.

As mentioned earlier, indirect prevention includes vaccination and sterilisation. Hepatitis B vaccination is now easily available from general medical practitioners and affords absolute protection against this condition once sero-conversion (that is, the production of antibodies) has occurred. Three inoculations are given, at 0, 1 and 6 months, and the preferred site is the deltoid muscle. Antibody levels should be measured prior to vaccination, again at the completion of the course, and at subsequent intervals, currently recommended to be 5 years. Should antibodies be present prior to vaccination, a course of treatment may not be necessary and expert advice should be sought.

Instruments and utensils (such as aspirator tips, glasses used for rinsing, and so on) may transmit organisms from patient to dentist, staff, or even to the next patient; steps must be taken to eliminate this route of transfer. Wherever possible, disposable items should be used, and used for one patient only. Subsequently, they should be disposed of carefully, preferably via an organised collection service, with the sharp materials in rigid containers and clinical waste in clearly marked bags, which will be incinerated. The Local Dental Committee should be able to advise on the facilities available in any particular area.

Care should be taken by the dentist not to put his saliva-contaminated or blood-coated, gloved hands on to areas that are difficult or impossible to clean. Picking out cotton wool rolls, adjusting the operating light or the dental chair and selecting instruments from a drawer are just a few examples of the potential for the transfer of organisms. Generally speaking, items used in the treatment and management of patients should be handled in the ways recommended in various authoritative publications, and it is important to establish a routine to clean handpieces and work surfaces quickly and effectively between patients. As registered practitioners, even though working as assistants, vocational trainees must act within guidelines considered to be reasonable, and must assume responsibility for the welfare of all those with whom they come into contact.

Radiographs are an essential part of diagnostic and procedural dentistry, and the effects of the ionising radiation produced during the taking of dental radiographs should not be forgotten. Indeed, such is the importance currently attached to the use and abuse of radiation that the GDP must now work within regulations and standards set by the authorities both from this country and from the EC. As a consequence of this, it is mandatory to register with the local Health and Safety Executive if radiographs are taken within your practice, as well as to organise an in-surgery radiation survey. This may be performed by any competent person in this field, or by a suitable organisation such as the National Radiological Protection Board.

Care should be taken not to cause unnecessary exposure of members of the dental team or the patient. Quality assurance is essential; not having to repeat a radiograph because of positioning faults, machine faults or developing faults and the use of good quality films and chemicals are important ways of reducing the amount of radiation a patient is exposed to.

Careful monitoring and adherence to international regulations will contribute towards good radiation hygiene, which will benefit the health of the patient, the staff and the practitioner. The importance of this has been underlined by the British Postgraduate Medical Federation, who, in their pioneering series of distance learning video tapes, have dedicated the whole of the second tape in the series to radiography, radiation and the current pertinent regulations.

Apart from transmissible diseases, it should be borne in mind that dentistry is both physically and mentally demanding. We are all guilty of adopting the most bizarre positions when treating our patients, and although it is done with the best of intentions for either seeing or doing something better, muscle fatigue and low back pain lie in store as a reward for this ungainly posturing. The reduction of physical stress to the body by the adoption of correct postures is an extremely important part of everyday practice and one that should be learned from a very early stage. Although low seated close support dentistry is now normal practice, there is still much scope for the abuse of muscles and joints.

Dentistry has a reputation for being a stressful occupation, particularly amongst dentists themselves. While any event within the surgery may act as a stressor (source of stress), anxious and uncooperative patients, the possibility of inflicting pain, and the pressure of keeping to a time schedule, are all common sources of stress. We appear to be prime candidates for 'burn out', a syndrome of emotional exhaustion and cynicism, frequently found amongst professionals who work directly with people. Although it may be difficult to prove such a situation, it is very important to recognise the stressors acting upon you (perhaps by keeping a 'stress log') and to attempt to change recurring patterns of stress which are producing anxiety. It is too easy to set unrealistic goals for ourselves and suffer emotionally in their non-achievement.

However, it is comforting to recognise that the stress which we experience during our working day is probably no greater than that of other health professions. Indeed, a certain amount of anxiety is essential for the production of peak performance; it is beyond the normal that anxiety becomes pathological and performance is impaired.

'After all is said and done, more is said than done!', and while it may be tempting to only pay lip service to the concept of patient, staff and dentist health within the surgery (perhaps on the grounds of time or cost), there is a moral, ethical and legal obligation upon the practitioner to care for all who pass through his surgery. Take care of your patients, take care of your staff and, most importantly, take care of yourself!

Bibliography

Mason R. *A guide to dental radiography*. Dental Practitioners Handbooks. Bristol: Wright (IOP Publishing), 1988.

British national formulary. London: British Medical Association and The Pharmaceutical Society of Great Britain, 1988.

Little J W, Falace D A. *Dental management of the medically compromised patient*. St Louis: C. V. Mosby, 1980.

Guide to blood borne viruses and the control of cross infection in dentistry. London: British Dental Association, 1987.

Scully C. *Patient care: a dental surgeon's guide*. London: British Dental Journal, 1989.

INCORRECT ASSUMPTIONS

One should be appreciated by all of one's patients

To be worthwhile, one must be thoroughly competent and successful in his field

One should become emotionally involved with one's patients

There is always a right, precise and perfect solution to a patient's problems, and this solution must always be found

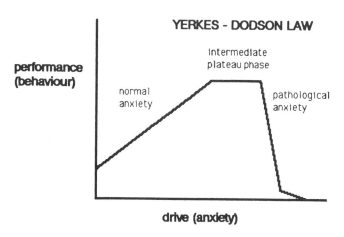

YERKES - DODSON LAW

performance (behaviour) — drive (anxiety)

normal anxiety — intermediate plateau phase — pathological anxiety

Infection control in general dental practice. Report of Symposium ICI Pharmaceuticals (UK), Alderley House, Alderley Park, Macclesfield, Cheshire SK10 4TF.

The ionising radiation (protection of persons undergoing medical examination and treatment) regulations, 1988. HMSO, 1988.

Stress in Health Professionals. Payne R, Firth-Cozens J (eds). Chichester: J. Wiley and Sons, 1987.

Useful addresses

Health and Safety Executive—Head Office, The Triad, Stanley Road, Bootle, Merseyside L20 3PG   Tel: 051 922 7211.

National Radiological Protection Board, Northern Centre, Hospital Lane, Cookridge, Leeds L16 6RW   Tel: 0532 679041.

12

Oral Surgery: The Diagnosis

R. L. Caplin

It is because we are experts in our field that patients seek our advice about the varied problems that they have in and around their mouths. In order to be able to give such advice, and to effectively treat the patient, a wide range of knowledge, skills and expertise has to be called upon by the practitioner. Listening to, observing, questioning, examining and performing further tests on the patient are all necessary stages along the pathway leading to a diagnosis. Only then is the practitioner in a position to consider the treatment options and to enter into discussions with the patient as to the best course of action to take. The time spent in this part of patient management is time well spent, and will greatly increase the likelihood of a successful outcome to the treatment.

Patients may present with a variety of problems, ranging from a small cavity to the acute emergency of a swollen face. Having been given advice about their condition, it usually follows that treatment is requested and is carried out at the same or subsequent visits, it being tacitly understood that once a patient has sought advice they will also wish to act upon that advice and have treatment performed.

A vast array of skills and knowledge are brought into play in arriving at a diagnosis; a practitioner listens to the complaints of the patient, observes them, questions them, examines them, performs further tests and then reaches a diagnosis and proposes treatment options. Making a diagnosis is imperative, and everyone should develop a protocol enabling them to elicit as much relevant information as possible from the patient in an efficient and sympathetic' manner. To this end, it is less threatening and far more relaxing to the patient if the initial consultation takes place away from the dental chair. This could be at a desk in the surgery (although not with the patient across the desk from you as this creates a psychological barrier), or in a quiet area somewhere in the practice, where the patient can feel that they have your full attention and where you can create some sort of rapport with them.

It is said that if you listen to the patient, they will tell you what is wrong with them, and clinical experience certainly bears this out. It is very important to listen carefully to what the patient has to tell you about their problem and to observe them while they speak, noticing any abnormality about their face, in their speech or in their behaviour. Having listened to the patient's story, a record must be made in the notes, using the patient's own words. This can then be followed by more specific questioning, in order to gain further information. For a routine dental inspection, this aspect can be dealt with very quickly and will consist of no more than an entry in the notes saying, for instance, 'complaining of a routine exam, no problems at present'. However, should the patient have a painful problem, a swelling or an ulcer, a more systematic approach will be needed. It is sensible to have a routine set of questions that are asked in a set order. In this

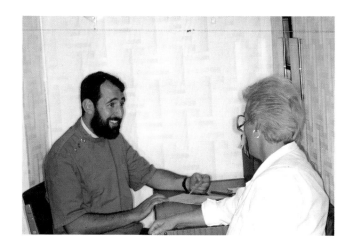

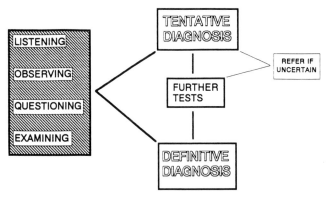

PAIN	Is it present at the moment?
	How long has it been present?
	Is it getting better or worse?
	What type of pain is it?
	(aching, throbbing, sharp, boring)
	Where is the pain?
	(localised to a tooth or an area or radiating)
	Is the pain affected by thermal changes?
	(hot or cold)
	Is the pain affected by sweet foods?
	Is there pain with pressure to a tooth or an area
	What relieves the pain? (analgesics, thermal changes, pressure changes)
	Are you kept awake at night?
	Is the pain less since the onset of the swelling?

| TEETH | Are any of them loose?
Have any broken recently?
Has there been any dental treatment
recently in the area? |

| SWELLING | When did the swelling appear?
Is it increasing in size?
Is the adjacent gum swollen?
Are there any areas of tingling or
numbness? (eg lower lip, tongue, cheek)
Is there a bad taste in the mouth and is it
coming from a particular area?
Is there any limitation of opening? |

| ULCER | How long has it been present?
Was it related to damaging the gum? (eg
tooth brushing, burning, food)
Do you get ulcers regularly?
How long do they last?
Do they subside by themselves?
Any restriction in movement of the area? |

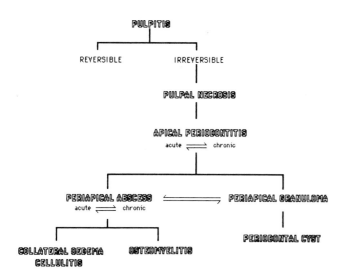

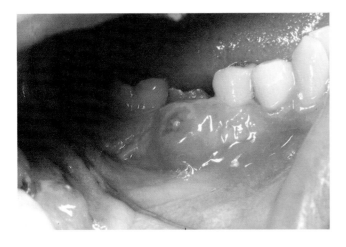

way, it is easier to match the patient's symptoms with the underlying disease process.

With a reversible pulpitis, the patient will experience bouts of sharp pain, lasting a few seconds and initiated by thermal changes. The progression to an irreversible pulpitis is accompanied by pain of a throbbing or nagging type, lasting at least several minutes, or sometimes hours, of spontaneous onset and worsened by thermal changes. Proprietary analgesics are often not effective, but do give the practitioner an idea of the severity of the pain and the patient's pain threshold. The acute pain of an irreversible pulpitis or an apical abscess will not respond for any appreciable length of time to 'over the counter' medications such as aspirin, paracetamol or any of their combinations with codeine, such as Paracodol or Codis. Irreversible pulpitis may persist for days or even months and some stoic patients resist seeking advice until the very last moment; for others, the pain may subside after a while. Unfortunately, although the patient may feel that the passing of the pain is a resolution of the condition, it is not. There will have been a progression to the next stage, which is death of the pulp (pulpal necrosis); this should be considered rather like an alarm clock ticking away nice and quietly until the alarm goes off (usually out of hours or while the patient is on holiday), and producing severe pain. An apical periodontitis ensues, the tooth becomes tender to touch or bite on and accurate localisation by the patient is possible. If, however, the site of the pain is vague, even though the pain itself is severe, pulp tissue is the most likely source of the problem.

With the onset of an apical abscess, severe acute pain will prevent sleep, diminish appetite and make the patient feel generally unwell. Should there be sweating or chills, resulting from an infection, the patient will experience considerable distress and anxiety. There is an intense, boring type of pain, as fluid builds up within the confined space of the apical periodontal tissues, and as the pus starts to track through the bony spaces. Usually, within a few hours or, at the most, days, the fluid pressure is released as it breaks through the cortical plates of bone to drain on to the oral or facial aspects of muscle bands, giving rise to a 'gum boil', a swollen sulcus or a swollen face in the majority of cases. Drainage on to the lingual aspect may give rise to a lingual 'gum boil' or swelling of the tissue spaces around the mandible, manifesting as swelling below or around the mandible. This is far less common than swellings of the face itself. The patient will now be at the stage of collateral oedema or cellulitis.

From the patient's history alone it is often possible to arrive at a tentative diagnosis before even looking in the mouth; the subsequent extra- and intra-oral examination merely confirms the diagnosis. At other times, however, the history and specific questioning may leave the clinician with several possibilities in mind and the subsequent examination must serve to reduce the possible diagnoses to a minimum. It is once again very important to establish a standard routine for the examination of patients, to be followed in every case, regardless of the apparent simplicity of the problem. In this way, signs will not be missed so easily!

The extra-oral examination begins with a general look at the patient's face, noting any asymmetry, alteration of skin colour or obvious abnormalities. Lymph nodes should be palpated (submental, submandibular and cervical) on both sides, to detect tenderness or enlargement. The intra-oral examination begins with a general look at the soft tissues, as far back as the fauces and the posterior wall of the pharynx (get the patient to say aagh, while depressing the tongue for a better view). The state of the tongue should be noted. If the patient has complained of a soft tissue problem, attention should now be focused on that particular area, for example, pain from an erupting wisdom tooth (pericoronitis), a lump on the tongue, an ulcer of the cheek or gum (aphthous ulcers) or a swelling related to a tooth (lateral periodontal abscess). Even if these are present, the examination does not stop there, but continues with a general survey of the teeth, restorations and appliances, becoming localised to the problem area as demonstrated by the patient. Any caries or periodontal problems in the immediate vicinity are noted, as well as any problems with restorations, teeth or appliances. Careful observations and their inclusion in the patient's notes are key elements in successful diagnosing.

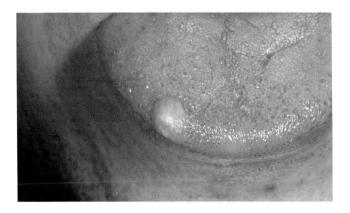

Having progressed from a general examination to a specific examination, the operator should have sufficient information to be able to reach a diagnosis or at least to have tentative diagnoses in mind. Further tests may then be carried out, with a view to confirmation or elimination of the possibilities. Remember that tests are for support of a clinical diagnosis, not for making the diagnosis. Radiographs, percussion, palpation, vitality tests, taking of the patient's temperature and biopsy are among the tests readily available to the GDP.

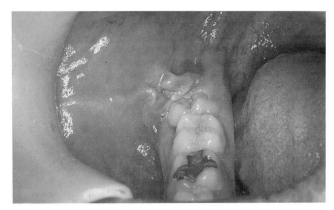

The most common further test in practice will be the radiograph, and it is clear that good clinical dentistry requires frequent taking of, and reference to, radiographs that are of good quality and show the relevant areas. Quality control in all its aspects is essential. For the vast majority of dental complaints, intra-oral views are adequate; the bitewing and periapical film provide a huge amount of relevant information. Other views, such as an occlusal or an oblique lateral are helpful in some circumstances and are well within the capabilities of the GDP, providing the correct radiographic equipment is available. A dental panoramic tomograph, if available, will also provide a large amount of information, but its limitations should be recognised, and they should always, where possible, be backed up with an intra-oral view.

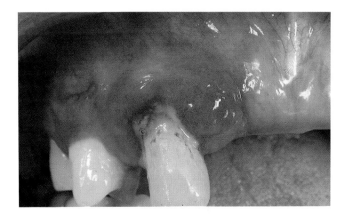

Soft tissue lesions are often diagnosed entirely clinically, but histological confirmation may be required or desirable, and it is often a great service to the patient to perform the biopsy in the surgery and send it to the local laboratory rather than commit the patient to much time loss by referring them to an oral surgery or oral medicine department. Performing these kinds of treatment is interesing and adds variety to the day. It does also, of course, enable a more reliable diagnosis to be made. The type of biopsy and the transport medium should be decided in consultation with the laboratory you are dealing with, and transportation of the sample to the laboratory should comply with any local or national regulations.

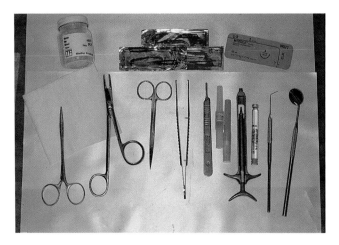

The uses of a thermometer in a dental surgery are limited, but at times essential, to enable the practitioner to assess the full extent and implications of the patient's condition. The presence of a raised temperature will most often be seen when the patient has an infection, giving rise to a cellulitis. In such circumstances, the prescribing of antibiotics would be

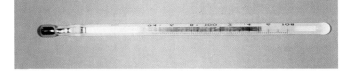

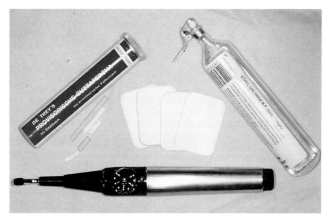

justified. A swollen face without a temperature does not require antibiotic therapy.

Percussion and palpation are very helpful in localising the responsible tooth, and in assessing the state of the periapical tissues. A tooth that is painful to tapping (periostitic) will have inflammation and increased fluid pressure at the periapical tissues; this is an apical periodontitis or apical abscess. Tenderness to palpation over the apex of the tooth, either buccally or palatally, is suggestive of tracking of fluid through the bone supporting the tooth, with resulting pressure to the cortical plate; this is an apical abscess that has spread.

When percussing a tooth, it is best to tap three teeth, starting with the one behind the suspect tooth, then the suspect tooth and then the tooth in front, giving each tooth a number and asking the patient to say which tooth was the most painful to the stimulus. The whole manoeuvre should be repeated, starting with a different tooth of the three and therefore giving them each a different number; that is, number one now becomes number three. In this way, the patient will have to give a genuine response and not an anticipated response and should repeatedly identify the painful tooth.

Vitality tests, using whichever method you find convenient and reliable, will give some information about the state of the pulp, but the limitations of these techniques should be recognised. Misleading results are common. The presence of fluid in a necrotic canal can give a positive result, and in multi-rooted teeth combinations of vital and non-vital canals will provide a variety of responses to vitality testing. With this test, as well as with percussion and palpation, it is useful to perform the test on equivalent teeth in the opposite arch in order to gauge the patient's response from normal teeth.

Once sufficient information has been accumulated about the patient and the presenting condition, it should be possible to make a definitive diagnosis, or at least to have narrowed down the possibilities. There may be several different ways of dealing with the problem and it is the practitioner's responsibility to advise the patient; this can only be correctly and honestly done if the diagnosis has been carefully arrived at. Once you know what is wrong, you can look at the treatment options and, in consultation with the patient, decide on a course of action.

Sometimes, of course, it may not be possible to arrive at a diagnosis, because of limited knowledge or expertise. In this situation, either seek guidance or refer the patient to someone who can help directly. It is not professionally demeaning to admit to not knowing the reason for a patient's suffering. It is, indeed most professional and certainly in the patient's best interests to ensure that he or she receives the best possible care, if not from you then from an expert in the relevant field. Patients respond very well to fallibility in a practitioner, if it is presented in the correct way.

It is essential to have a systematic and rational approach to a patient's complaints; there can only be one correct diagnosis and every effort must be made to reach it.

13

Oral Surgery: Assessment and Treatment

R. L. Caplin

In arriving at a diagnosis, a dentist will have employed much skill and expertise, but there can be no doubt that in formulating a treatment plan, in addition to skill and expertise, intuition and wisdom (and sometimes inspiration) are required. Whereas the road to diagnosis is paved with objectivity, the road to treatment is paved with subjectivity and instinct.

We are required, as providers of health care, to decide what is in the best interests of our patients and what treatments will best deal with the problems they present. Not that the patients are excluded from this process; they should be, indeed must be, involved in the decision making process, and we need to know their view of their condition and its management by offering them the treatment options available. We cannot, however, abrogate our responsibility, and it is with us, the experts, the professionals, that the final decision regarding treatment rests.

Assessment subsequent to diagnosis is, therefore, a crucial stage in the process of healing the patient. It is a correlation of as much gathered information as is necessary to formulate a treatment plan which will, with the minimum disruption, destruction or violation of the patient, either physically or emotionally, render the patient symptom-free and healthy. Local and general, physical and emotional factors all have to be considered.

For example, in the case of an irreversible pulpitis resulting from caries, it is generally accepted that treatment must involve removal of the offending pulpal tissue. This can be achieved by either removing the pulp without the tooth, that is, root canal therapy, or removing the pulp with the tooth, that is, extraction. Clearly, the blind dogma of following one particular treatment modality in a situation of this type is not in the best interests of the patient.

From a local point of view, is root canal therapy tenable? Is there a good prognosis? Is access to the canals and to the apices favourable? Can the tooth be adequately restored after removal of the caries and tooth tissue during root canal therapy? Generally, does the state of the dentition warrant retention of this tooth? Are there several other teeth missing that will require replacement? Can the tooth (if removed) be added to, or included in, a denture? What is the periodontal status? Physically, is removal likely to be straightforward or require surgery? Are there adjacent vital structures to beware of in this eventuality? Emotionally, how much treatment can the patient tolerate in his or her present state? Is sedation or general anaesthesia required? Can the patient make a reasonable decision in his or her present state, or will he or she be looking for immediate pain relief rather than long-term benefits? Can the patient tolerate long periods of treatment in the chair? Can he or she afford the cost of treatment?

Furthermore, in formulating a treatment plan, the practitioner will not only be taking these factors into account, but will also need to assess his or her own ability to cope with the situation. If retention of the tooth is favoured, but the dentist feels unable to manage the treatment, should he refer the patient? The same applies for an extraction. Alternatively,

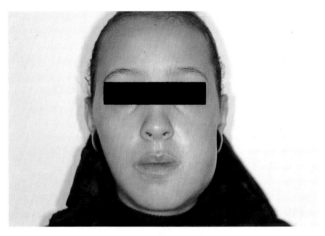

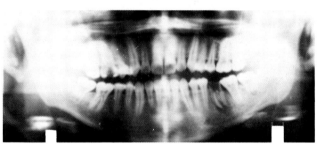

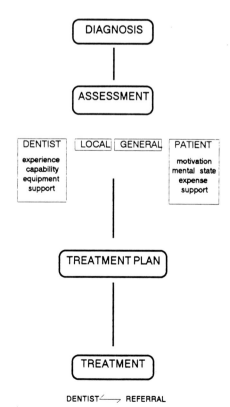

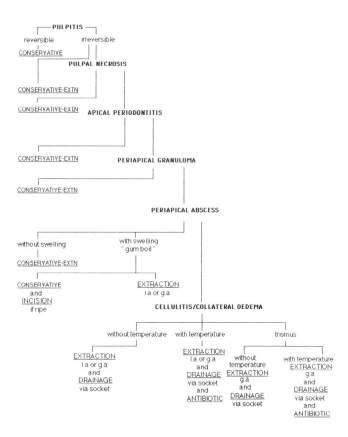

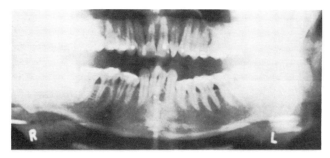

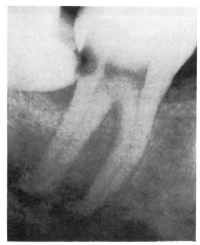
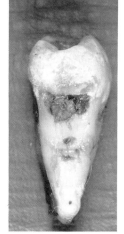

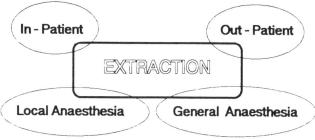

should the treatment plan be confined to the dentist's own perspectives and capabilities?

These are matters of fine judgement, but it behoves the dentist to arrive at a treatment plan that considers all of the above factors and then to place before the patient the reasonable treatment options, even if these are beyond his personal skills. A patient involved in the decision making process will be a much more sympathetic and cooperative patient. This is not advocating reneging on our professional responsibilities and allowing the patient to direct the course of treatment; it is advocating the need for the informed consent of the patient. Patients are entitled to be made aware of the possible courses of action, and their respective merits and pitfalls. Under no circumstances, however, should a practitioner be persuaded to undertake a course of action with which he is not completely sympathetic.

This is clearly illustrated with regard to a patient suffering from pericoronitis. There are three clinical gradations of this condition: acute, subacute and chronic, and all are related to a third molar (usually lower) in a state of partial eruption. Management will depend on the diagnosis and assessment of the situation. Chronic pericoronitis is self-limiting, of little consequence to the patient, and does not require active intervention, irrespective of the state of eruption of the associated tooth. Subacute pericoronitis manifests as a bout of well localised pain, constant in nature, often worse with swallowing and with a swollen gum flap over the tooth, from which emanates a bad taste.

While causing discomfort, this, the most commonly seen type of pericoronitis in practice, is also self-limiting, and requires little more than gentle irrigation of the flap and perhaps elimination of trauma from an opposing tooth by grinding of the offending cusp. The patient himself can manage the situation further with hot salt mouthwashes. Again, treatment is irrespective of the state of eruption of the associated wisdom tooth and extraction is not indicated unless the bouts of subacute pericoronitis recur often enough and are debilitating enough for the risks of post-extraction complications to be outweighed by the disruption of the patient's life.

Acute pericoronitis (swollen gum, trismus, pyrexia, severe pain), however, requires immediate and active intervention with antibiotics, drainage and irrigation, often within a hospital environment. Such a condition is a strong indication for removal of the offending tooth.

Thus, having arrived at a diagnosis, it is necessary to have definite guidelines for removal of a wisdom tooth. These should be (1) untreatable caries in the tooth itself or the adjacent tooth, (2) repeated and frequent bouts of subacute pericoronitis or one attack of acute pericoronitis, or (3) an enlarged perifollicular space. The BIT (Because it's There) syndrome is not an indication for removal, nor is the fact that the fee structure within the NHS makes efficient removal of wisdom teeth lucrative.

The assessment of wisdom teeth is well covered in standard textbooks of oral surgery, but the operator has to know his own limitations and capabilities once a decision has been made to remove the tooth.

The operator will have to judge whether the patient can cope with the removal of the tooth under local anaesthesia, or whether general anaesthesia would be preferred or desirable? If general anaesthesia is decided upon, the remaining wisdom teeth should be assessed for removal under the same anaesthetic. Is sedation or general anaesthesia in the surgery available? If so, is there an experienced anaesthetist or trained dentist available to administer the drugs? Is resuscitation equipment available if needed, and are there adequate recovery facilities?

It is known that recent graduates are inexperienced in the surgical removal of teeth and roots, especially wisdom teeth, and will need the close support and tuition of a competent practitioner (as in the vocational training scheme), or practice within an oral surgery department (assistantships can be arranged with local consultants) in order to gain the necessary experience and confidence.

It is very easy and tempting to exclude this aspect of treatment from your repertoire, but in doing so, you are not offering your patients a complete service. However unpleasant the prospect of an extraction may be, a patient who has confidence in his dentist will often prefer the procedure performed in the familiar surroundings of his general practice, in a quick and efficient manner, rather than spend time travelling and waiting for hospital advice, followed perhaps by months of waiting for treatment.

As far as the assessment of the tooth is concerned, removal should not take more than 30 minutes from administration of the local anaesthetic to dismissal of the patient. If this seems a short amount of time for an inexperienced practitioner, it is! However, it is a useful guide; beyond the 30-minute barrier, both the dentist's and the patient's temper will be pushed to the limit. For surgical removals estimated to take longer than this, the patient is best treated under general anaesthesia.

Adequate equipment must be available; this should include a range of forceps and elevators, bone removing equipment (that is, slow handpiece and surgical burs) and suitable irrigant. A regular contra-angle handpiece can be used, provided it is sterile, and sterile saline (available from pharmacies) can be delivered to the operative site by a chip or disposable syringe. An air rotor must never be used for bone removal, because of the risk of surgical emphysema. Suture material must also be available, as well as haemostatic agents (for example, Surgicel). In trying to deliver the tooth, it is often profitable to try gentle elevation before raising a flap, as this will often result in removal of the tooth without the need to open tissue planes.

It is sensible to warn the patient of the post-operative complications that may occur, particularly swelling and pain. If you have warned the patient about swelling and none occurs, you will be credited with having done a good job; on the other hand, failure to warn, followed by swelling, however reasonable, will be considered by the patient to be your fault. It is also worthwhile suggesting an analgesic, as there will certainly be post-operative pain.

As part of total patient care, make contact with the patient over the next day or so, to make sure that all is well and to sort out any problems that may arise. Patients are often reluctant to disturb a dentist, even if they are in considerable discomfort, and so it should be up to the dentist to establish lines of communication with the patient. If the patient were in hospital, apart from constant supervision by the nursing staff, he or she would be seen twice daily by members of the dental team. Similar contact in general practice gives the patients confidence, knowing they can relate their problems without

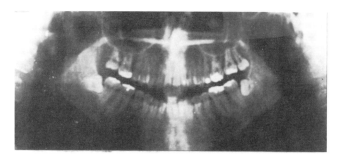

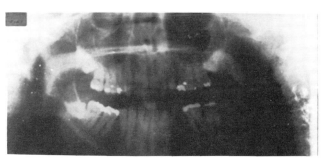

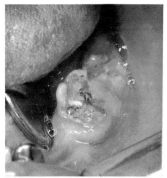

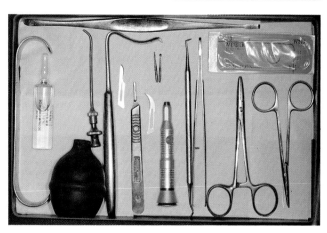

1. Do not eat or drink on the side of the extraction. When eating or drinking use the other side of the mouth.

2. After 24 hours rinse the affected side with hot salt mouth washes (a tablespoonfull of salt in a glass of almost boiling water) and continue every 4-6 hours until the soreness has gone.

3. Should the socket start to bleed again roll a handkerchief into a ball and place it on the site of the bleeding and bite for 20 minutes.

4. If you have pain from the extraction site after a few days please contact your dentist for advice.

5. If your have had lower wisdom teeth out you can expect some swelling and soreness in the region. You may also be unable to open your mouth to its full extent. This should subside within 4-7 days.

3A DENNIS PARADE,   886 3355   R. L. CAPLIN
WINCHMORE HILL ROAD, N14 6AA

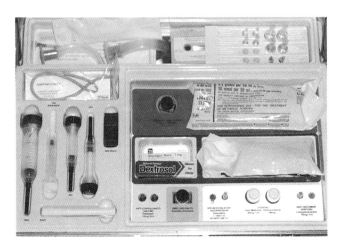

feeling as if they were intruding upon the dentist's time.

Good back up and support are essential for the efficient and effective surgical removal of teeth. At the chairside, support from a dental surgery assistant in retracting tissues, handing over instruments and irrigating the operative site will enable the operator to concentrate on the task at hand and not divert his attention to peripheral matters. In addition, it is vital for the inexperienced operator to have advice and help available from a colleague should the procedure not go according to plan and problems arise.

Further factors to be taken into account when assessing a tooth are the time of day scheduled for its removal and whether or not, after the extraction, the patient is going to be away from dental support should it be needed (that is, going on holiday, travelling to remote places). It is unwise to surgically remove a tooth late in the day; should post-extraction complications arise (for example, haemorrhage), the patient may be forced to seek help at 8 or 9 o'clock at night, or later. If you are called out, and your number should be available to the patient, this will be quite a disruption. Should the patient attend a local casualty department, the treatment he or she receives at the hands of an inexperienced casualty officer (medically trained) may not be in anybody's best interests. Similarly, it is best to postpone an extraction until such time as the patient will be able to be around, following the removal, to seek advice should it be necessary.

The patient's ability to cope with a surgical removal must be taken into account when deciding when, where and how the tooth is to be removed. If unable, or unwilling, to have local anaesthesia, sedation or general anaesthesia could be offered, but only if adminstered by an experienced person, and not with the operator acting as the anaesthetist. Experience in these areas can be obtained by attending courses (for example, SAAD), or by undertaking an assistantship in anaesthesia, which can be arranged initially with the local oral surgery consultant and/or the anaesthetic department of a dental hospital.

It is a *sine qua non* that adequate resuscitation equipment is available at all times, whatever method of anaesthesia is used, and that all staff are trained in, and conversant with, resuscitation procedures. Regular training sessions should be held within the practice, and the local branch of The British Red Cross or St John's Ambulance Brigade will be able to assist in this respect.

Assessment is thus an extremely important part of the chain in successfully treating a patient with problems, and it is honesty, integrity and professionalism that sets apart the good dentist from his colleagues.

References

Seward G, Harris M, McGowan D. *Outline of oral surgery*, Part 1. 2nd ed. Dental Practitioner Handbooks. Bristol: Wright, 1988

MacGregor A J. *The impacted lower wisdom tooth*. Oxford: Oxford University Press, 1985

14

Oral Surgery: Problems

R. L. Caplin

Having listened to, examined, tested, diagnosed and recommended a course of action to the patient, treatment will usually follow at the same or subsequent visits. In the vast majority of cases, the foregoing will have proceeded without a hitch, and the treatment completed to the satisfaction of both the practitioner and the patient. Years of study, and the acquisition of knowledge, skills and expertise, will have been condensed into a short amount of clinical time. Unfortunately, this highly desirable state of affairs cannot apply every time; problems can arise anywhere along the pathway leading to the successful completion of treatment. Some patients will present difficulties in diagnosis and treatment planning, whereas others will have problems either during the execution of treatment or post-operatively. Whenever a difficulty arises, it is essential to identify the problem, reappraise the situation and decide upon an appropriate course of action; flexibility is the key to success! In this chapter, three difficult areas are looked at: the fractured root, the use of antibiotics and the swollen face.

A relatively common problem that arises during the removal of teeth is the fracture of the tooth, leaving a remnant of varying size within the socket. A decision then has to be made regarding this retained fragment, namely whether to remove it or leave it *in situ*. Clearly, the operator should have definite guidelines upon which to base such a decision. The original reason for the removal of the tooth is an essential part of the consideration of the future of the retained fragment. Where a tooth is being removed because of periapical disease (such as apical periodontitis or apical abscess), failure to remove the apical part of the tooth will in fact be leaving behind the very part involved in the disease process and responsible for the patient's symptoms. This can only lead to a prolongation or intensification of the problem. In such cases, the root fragments must be removed, in order to be certain of eliminating the disease process.

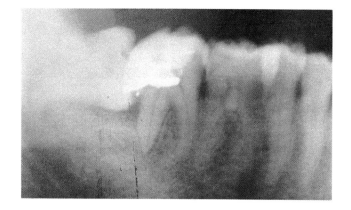

The size of the remaining fragment is also an important consideration. Arbitrarily, where more than 3 mm of the tooth remains, the fragment should be removed; less than this, with other factors being favourable, the root can be left, kept under observation and the patient informed.

The stage during the actual extraction at which the fracture occurs is another significant factor. Sometimes, no sooner will forceps have been placed on the tooth and rocking movements commenced than a crack will be heard, indicating that fracture has taken place. The tooth may still be firm, and when it is eventually removed the remaining fragment will be firmly retained within the socket. On the other hand, after the application of forceps and appropriate movements, the tooth may be very loose and in the final stages of delivery when the fracture actually occurs; the tooth is again delivered devoid of the fractured part. The retained fragment is quite loose within the socket. In this situation, the mobile root, having been separated from its periodontal support, may be considered a 'foreign body' and must be completely removed. The firmly retained root is still as it was before the extraction and again, all factors being favourable, can be left under observation.

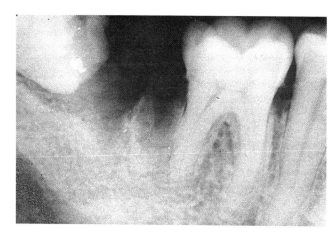

A further consideration is the variable difficulty and

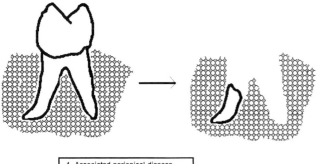

1. Associated periapical disease.
2. Size of the remaining fragment.
3. Mobility of the remaining fragment.
4. Difficulty in removing the fragment.
5. Subsequent orthodontic treatment.

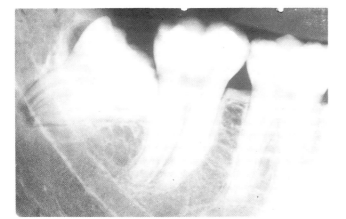

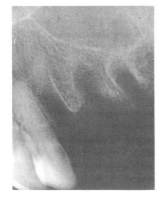

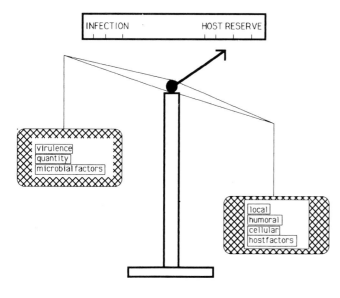

destruction involved in removing the retained fragment and the ability of the operator to effect its removal. The removal of a partially erupted lower third molar which has been responsible for bouts of pericoronitis may be accompanied by the fracture of the apices. Removal of these fragments could require the raising of a flap, extensive removal of bone and all the post-operative complications that can accompany such procedures. If the retained apices were not originally involved in a disease process, the disadvantages of retaining the roots must be weighed against the disadvantages of attempting to remove them. In many instances, discretion is the better part of valour; the retention of these roots is not usually associated with post-operative problems.

Adjacent vital structures should also be taken into consideration, for example the proximity of the maxillary antrum in relation to retained roots from the upper first and second molars (especially the large conical palatal roots). These may be easily forced into the antrum when inappropriate elevation is applied.

Proximity of roots to the inferior dental canal and the mental foramen should also be carefully observed. Prolonged, or permanent, post-operative paraesthesia or anaesthesia are real risks. No patient will thank you for such an outcome, and some will be less understanding than others.

Immediately following the removal of a tooth, it should be inspected to ensure that it has been removed in its entirety. Suspicion should be aroused if the apices feel sharp when rubbed with a finger; natural processes generally result in rounded surfaces, so that a shortened, sharp-ended root has almost certainly been fractured. A radiograph should always be taken to localise accurately the retained fragment and its proximity to vital structures. This radiograph will also serve as a baseline for future reference. Finally, if removal of the tooth is part of an orthodontic treatment plan, it is preferable not to leave fragments behind as they may interfere with proposed tooth movements. In any situation where a tooth has fractured, the patient must be informed and an appropriate entry made in the notes.

Another 'grey area' is the use of antibiotics in the treatment and management of patients. In a healthy individual, there is a balance between the resident microbial flora and the host's defence mechanisms. An alteration in one will lead to an alteration in the other, producing a clinical or subclinical condition. The alteration may therefore be a quantitative one (that is, a change in the number of organisms) or a qualitative one (that is, a change in the type or virulence of organisms), or could be due to an alteration in the body's defence mechanisms. AIDS is a recent example of this, as is immunosuppression as a result of steroid therapy.

However, for the vast majority of the cases seen in dentistry, any imbalances are entirely localised and are kept that way by the body's ability to mount an effective response to the situation. Where there is an impasse between pathogenic organisms and defence mechanisms, a chronic situation develops.

A common example of this is an acute periapical reaction, arising from an infected necrotic pulp. The pulp may be seen as the seat of the infection, organisms having arrived as part of the carious process or by direct communication with the mouth. Within the root canal, the organisms, even if bathed in tissue fluid that has exuded in through the periapical region into the canal, are inaccessible to the normal defence mechanisms of cells and antibodies. This undisturbed production line of organisms (probably anaerobic) leads to a local quantitative imbalance, with the organisms either proliferating into the periapical region or releasing destructive chemicals there. This imbalance in the periapical region will provoke a response which, because of the quantity of organisms and their continuous production, will probably take the form of a walling off of the area, that is, a chronic inflammatory response. This may be seen as a periapical granuloma, perhaps visible on a radiograph as a periapical radiolucency. Practices where rotational tomographs are routinely taken are frequently presented with the dilemma of asymptomatic periapical radiolucencies; their management depends to a large extent on the philosophy of the individual practitioner.

This state of affairs may exist for years, producing mild if any symptoms for the patient. However, a sudden further change, either in the number, virulence or type of organism, can alter the status quo and produce an acute periapical abscess. Clinical experience shows that the imbalance thus produced can be restored to normal if the source of the organisms is removed; this means removal of the contents of the root canal, either by means of an extraction or through root canal therapy. In a healthy individual, able to mount a normal response, this will allow a return to normal. As in military strategy, if the supply lines are cut off, the advanced troops are unable to cope as well and are far more likely to be overwhelmed.

Thus, local mechanical measures, where appropriate, are the most desirable means of resolving oral conditions, since they are the least invasive and least potentially harmful to the patient. Where there is pus or an accumulation of tissue fluid, it must be drained!

However, having mentioned that most oral infections are localised and best treated without the use of antibiotics, if an infective process is suspected, it is wise to record the patient's temperature; a rise from the normal is strongly suggestive of systemic involvement (that is, the organisms are no longer contained locally). In this situation only, it is recommended that antimicrobial therapy be instituted, in addition to local mechanical measures.

It is also clear that for some conditions (streptococcal, candidal or herpes virus infections), antibiotic, antifungal or antiviral therapy is required. Further difficulties then arise over the choice of antibiotic and the best regimen.

Most infections in the mouth are 'mixed' or polymicrobial, often involving Gram-positive aerobic and anaerobic organisms. In the absence of sensitivity tests, the practitioner has to prescribe according to predetermined guidelines and may well find that, in the absence of a reasonable improvement, the drug prescribed is either ineffective against the particular causative organism or that the dosage itself is inadequate. Since accurate diagnostic cultures may require a week or more to produce a reliable result, treatment has to begin based on circumstantial evidence. It is generally preferable to use higher doses for shorter periods of time. This should result in a subjective feeling of improvement for the patient within 24–36 hours. It is important to realise that the risk of allergic response to an antibiotic is not dose-dependent. Failure to show some measure of improvement

CONDITIONS REQUIRING ANTIMICROBIALS

Acute pericoronitis

Fascial space infections with an associated pyrexia

Candidal infections

Viral infections

Acute maxillary sinusitis

FAILURE OF TREATMENT OF INFECTION

1. Failure to institute adequate local measures eg drainage of pus, irrigation of a gum flap

2. Poor host response eg underlying systemic disease

3. Failure of antibiotic to reach the site eg osteomyelitis

4. Inadequate antibiotic dosage

5. Clinically wrong bacteriological diagnosis

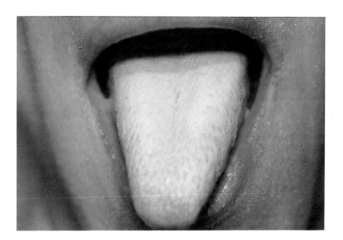

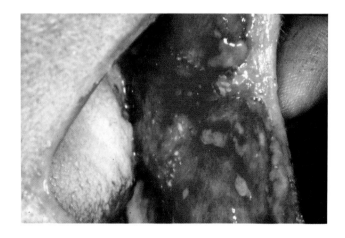

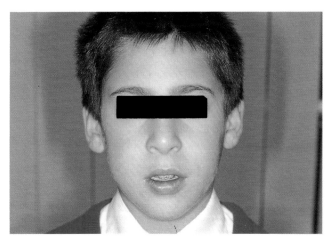

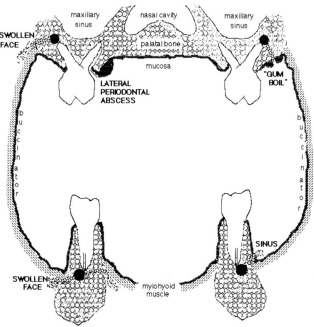

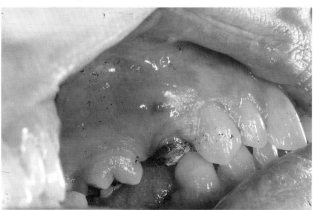

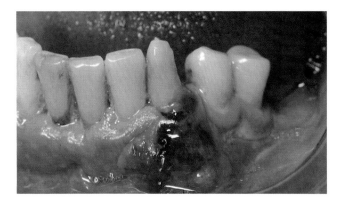

within this time necessitates a reappraisal of the condition and its management.

In principle, a practitioner faced with a patient who has a dental infection should think about reaching for his instruments first and prescription pad and pen later, if at all.

Another area of difficulty lies with the patient who presents at the surgery with a swollen face. Since about 72% of swollen faces are of dental origin, it is evident that the GDP has an important role to play in the treatment and management of these problems.

Any distortion of the face, which is our main means of non-verbal communication, can have an alarming effect upon not only the individual involved, but on the family and even on the practitioner asked to see the patient. However alarming the presentation may seem, a clear understanding of the mechanisms involved will once again enable the dentist to adopt a rational approach and deal efficiently, confidently and effectively with the situation.

Upon presentation, the dentist will have to elicit a full history from the patient (see chapter 12), perform a thorough extra- and intra-oral examination, carry out any further investigations required and then arrive at a diagnosis. A treatment plan must then follow.

There can be a wide variation in the history leading to a swollen face. On the one hand, the patient may have experienced the extreme pain of an irreversible pulpitis, followed within hours by the appearance of a swollen face. On the other hand, a non-vital tooth may have been asymptomatic for years when, without warning or apparent cause, an acute apical abscess develops, to be followed by a swollen face. In most cases of sudden onset of a swollen face of dental origin, the patient will give a history of toothache. This is well localised to an area of the mouth or to a particular tooth, is getting worse and keeps the patient awake at night. The tooth is tender to touch.

Usually, following death of the pulp (for example, through caries or trauma) there is a breakdown of the pulpal cells and the release of their intracellular chemicals into the periapical region, resulting in an apical periodontitis. This may or may not be accompanied by the passage of organisms, and even though organisms may be isolated from the perapical tissues, this cannot necessarily be described as an infective process. At this stage, the patient will be able to localise the pain accurately, compared with the vague localisation typical as pain of pulpal origin.

The apical periodontitis may lead to an apical abscess, which is an accumulation of pus within the periapical tissues. The patient will now have a severe, sharp, throbbing pain, localised to an exquisitely tender tooth.

Generally, the continued production of pus can either produce pressure necrosis, resorption, or even the chemical stimulation of bone. The pus often tracks through the medullary bone, along the line of least resistance, producing an excruciating 'boring' type of pain for the patient.

Fortunately, the pus tends to perforate the buccal or lingual plates rather than accumulate within the medullary bone, and the release of the fluid and the reduction in pressure that this perforation brings is usually accompanied by a reduction in the level and type of pain. The 'boring' pain becomes a dull ache and the patient begins to feel better.

The combination of severe pain, lack of sleep and lack of

appetite that accompany this condition, will leave the patient feeling weak and rather unwell. A further complication may be the presence of a fever (with sweating and or chills); this strongly suggests that the infection is not localised to an area or tooth, but is now systemic. The next stage will depend on the anatomical relationships.

In almost similar circumstances, a disease process which gives rise to a facial or buccal swelling as a result of the drainage of tissue fluid could, alternatively, give rise to a buccal sinus or a 'gum boil'. The determining factor in the outcome is the relationship of the apex or apices of the involved tooth or teeth with the adjacent muscles and their insertions. In facial swellings, the muscle most frequently involved is the buccinator.

Where an apex is attached above (in the upper arch) or below (in the lower arch) the insertion of the buccinator, any inflammatory exudate, tissue fluid or pus produced will track along the line of least resistance, passing through the buccal plate of bone, to be released on the facial side of the muscle and producing a facial swelling. Such a swelling is termed collateral oedema and is distinguished from a spreading infection (cellulitis) by the normal colour and temperature of the skin, as opposed to the red and hot skin of a cellulitis.

Alternatively, where the fluid drains on to the oral side of the muscle, because of the apical relationship of the muscle, a sinus will develop (although not necessarily directly over the affected apex of the tooth) allowing drainage of the fluid into the mouth. When drainage is obstructed, either a 'gum boil' will develop or a localised swelling of the buccal sulcus.

Should drainage occur close to a muscle of mastication (either the masseter or the anterior fibres of the temporalis, where they extend down the anterior border of the external oblique ridge just distal to the lower third molar), limitation of opening (trismus) may follow and severely restrict the clinical examination and treatment options at this stage. When an upper canine is involved, swelling will be seen at the lateral aspect of the nose, obliterating the fold between the nose, lip and cheek and producing a swelling at the inner aspect of the eye and the lower lid.

It is important to remember that, whatever the degree of swelling, there is generally little to be gained by incision to achieve drainage unless the swelling is 'ripe', that is, well circumscribed and tense, rather like a balloon filled with water and ready to burst. Many of us have incised buccal swellings, only to fail to achieve any drainage.

With the onset of the facial swelling, often occurring overnight, the pain will have reduced, if not completely subsided. There may be some intra-oral swelling in addition to the facial swelling.

Examination will reveal a swollen face, with or without lymphadenopathy (swelling and/or tenderness of the draining lymph nodes), the texture of the swelling being soft in the case of collateral oedema and firm in the case of a cellulitis. The patient's temperature may be raised, hence the importance of having and using a thermometer. The combination of a pyrexia and regional lymphadenopathy is the only indication of the use of antibiotics in addition to local measures. Trismus will be observed where the masseter or temporalis muscles have been involved, and paraesthesia of the lip or cheek where the swelling is close to a bony foramen through which a nerve exits from the skull (that is, mental and infra-

SITE OF SWELLING	TEETH INVOLVED
upper lip (due to swelling in the labial sulcus)	upper 12
nasolabial fold, inner aspect of the eye and lower lid (giving the appearance of a closed eye)	upper 3
cheek (in the region of the infraorbital foramen)	upper 45
cheek, tending to be larger about the body of the mandible (buccal space) NOTE: do not assume that a swelling at this level has arisen from a lower tooth; a swelling arising from an upper tooth will drain under the influence of gravity	upper 678 lower 345678
lower lip and chin (the swelling of the face is usually the result of a swollen buccal sulcus)	lower 12

SIGN	TEETH INVOLVED
limitation of opening	lower 78
numbness or tingling of lower lip	lower 45
numbness or tingling of cheek	upper 3456

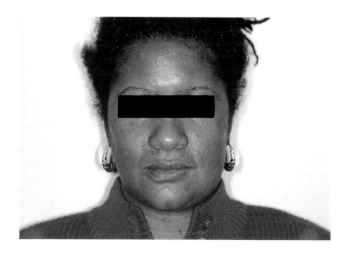

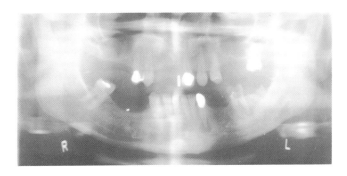

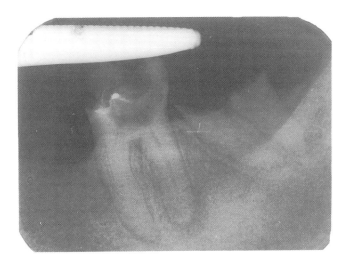

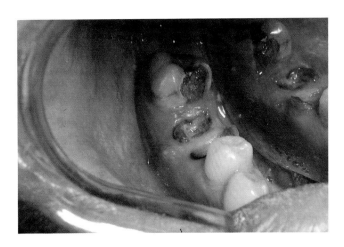

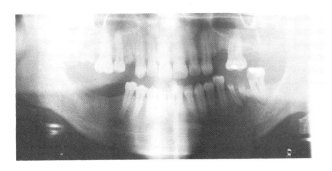

orbital nerves). Closure of an eye will follow involvement of an upper canine.

Radiographs will confirm the involved tooth or teeth and a periapical is the view of choice to show the number of roots present, their angulation, bone levels and the proximity of important structures such as the inferior dental canal, the mental foramen and the maxillary sinus. Where trismus precludes such a view, then an extra-oral view such as an oblique lateral or a dental panoramic tomograph will be adequate. A thermometer will demonstrate the presence of a pyrexia and vitality tests should reveal the state of the pulps of the teeth in the affected area.

Although there can only be one correct diagnosis, and the principles of treatment will always be the same, there may be different ways of managing the problem.

For these patients there must always be drainage of any accumulated fluid and elimination of the tissue which is the seat of the disease process. These objectives may be achieved by either removal of the offending tooth or teeth, or removal of the pulpal tissue by root canal therapy. A swollen face represents a dental emergency and careful consideration should be given to the form of treatment undertaken. Drainage is best achieved through the socket following extraction and only in exceptional cases should root canal therapy be considered.

Despite the inherited wisdom that such situations cannot be managed under local anaesthesia, because of the risk of 'spreading the infection', clinical experience does not support this premise. For the vast majority of cases, removal of offending teeth can be performed with local anaesthesia and without the risk of compromising the patient.

In the lower arch, inferior dental, long buccal and lingual nerve block injections will provide an adequate depth of operating anaesthesia, without the need to deposit solution close to any inflamed or infected tissues. In the upper arch, to avoid such tissues, the local anaesthetic is deposited in larger quantities than usual (2·2 ml plus), both at the mesial and distal limits of the intra-oral swelling. The needle is inserted approximately two teeth beyond the swelling and pointed towards the apex of the tooth. For the posterior infiltration, the needle is bent to about a 45 degree angle, again to allow insertion two teeth behind the swelling and to be pointed towards the apex of the involved tooth. A normal palatal infiltration will be effective. About 15 minutes should be allowed for the local anaesthetic to diffuse through the tissues before attempting the extraction.

Such teeth are generally easier to extract; they are often periodontally involved and the associated disease process will have produced further bony support loss It should be remembered that patients in this situation will have been suffering from severe pain and lack of sleep, and will not have eaten properly for a while, all contributing to a lowered tolerance of treatment and threshold of pain. Thus, it may not be possible to achieve a completely painless and pressure-free extraction, but this should not deter you from performing the required treatment immediately.

Having delivered the tooth, drainage may begin immediately through the socket and can be seen as a watery fluid mingling with the oozing blood. Thicker pus may also be seen. Drainage is encouraged by placing a finger in the buccal sulcus and gently massaging the tissues by moving the finger at the level of the buccal sulcus towards the extraction site. This is performed both mesially and distally to the socket, commencing at the limits of the swelling. The swollen face is also gently massaged and the socket observed for further drainage. Eventually, the bleeding from the socket will be the normal bluish red colouration, indicating that the fluid cavity has been drained and that bleeding is now coming directly from the highly vascular cavity wall. Although the swollen tissues may collapse initially, they will fill again with the oozing blood. Reduction of the swelling will begin to take place within 24–36 hours, but there will be earlier relief of the symptoms.

Antibiotic therapy is not indicated in these situations, unless there is an associated pyrexia. There is no need to give antibiotics 'just in case'. Most swellings generally do not yield sufficient quantities of organisms to produce cultures and cannot therefore be considered as infective processes.

Even if it is considered inappropriate to extract the tooth, perhaps because of lack of confidence by the practitioner or poor access due to trismus, the treatment of choice is nevertheless removal of the tooth. If achieving this means referral to an oral surgery department for extraction under general anaesthesia, either as an out-patient or an in-patient, then this should be done as soon as possible.

Such conditions are emergencies and in some circumstances, especially in the lower jaw, spread may occur into the tissue spaces adjacent to the tongue (submandibular and sublingual spaces) and may become life threatening.

Following treatment, it is advisable to see the patient again within 24 hours, in order to ensure that the situation is not worsening and that some improvement is taking place. Indeed, contact should be maintained with the patient until resolution has taken place.

Problems will inevitably arise during the many and varied treatments that we carry out. Recognising the problems and taking appropriate action is part of being a professional. If something goes wrong, this is not necessarily negligence, professional misconduct or a breach of Terms of Service; failure to recognise the situation and attempt to put it right may be.

Further reading

1 Pallasch T. J. *Can Dent Assoc J* May 1986.
2 Topazian R G, Goldberg M H. *Management of infections of the oral and maxillofacial regions.* Eastbourne: W B Saunders, 1981.
3 Barker G R, Qualtrough A J E. An investigation into antibiotic prescribing at a dental teaching hospital. *Br Dent J* 1987; **162:** 303–306.
4 Birn H. Spread of dental infection. *Dent Pract* 1972; **22:** 347–356.
5 Heimdhal A, Nord C E. Treatment of orofacial infections of odontogenic origin. *Scand J Infect Dis* (Suppl) 1985; **46:** 101–105.
6 Labriola J D, Mascaro J, Alpert B. The microbiologic flora of orofacial abscesses. *J Oral Maxillofac Surg* 1983; **41:** 711–714.
7 Lewis M O. Acute dentoalveolar abscess; microbiological and clinical studies. PhD Thesis, University of Glasgow, 1987.
8 Schein B. Microbiological considerations in selecting a drug for endodontic abscesses. *J Endod* 1986; **12:** 570–572.

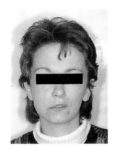
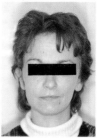
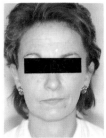

Pre LA extraction   At 6 and 24 hours—no antibiotics

15

Prosthetic Dentistry: Technical Support

R. D. Welfare and S. M. Wright

Prosthetic dentistry in general practice produces many challenges for both the newly-qualified trainee and the long-standing practitioner. If these challenges are not appreciated, met and dealt with effectively, the dentist may be left disillusioned with his skills and expertise, and stressed at the thought of treating patients requiring dentures. The economic need for rapid treatment while maintaining good standards of work can result in pressure, and the temptation to 'cut corners' can prove costly. These two chapers, far from being a definitive text, seek to illuminate some potential problems and to lead the clinician to establish his own solutions in the areas of the laboratory, the prescription, communication and some alternative clinical techniques.

For most trainees the thought of having to organise and arrange laboratory facilities for themselves for the first time is quite daunting. Fortunately, however, in most instances, they will probably use the practice laboratory to start with. Although an easy option, they should not miss the opportunity to understand the selection processes which need to be followed to set up a system which is effective and provides the necessary technical support for their requirements.

One of the most useful activities is to visit the laboratory. Hopefully, the principal will introduce the trainee to the technicians, so both can put a face to the voice at the end of the telephone and the trainee can see the laboratory facilities at first hand. If the principal does not arrange this, the trainee should do it him/herself. This is worthwhile, even if a postal system is used and the laboratory is some distance from the practice. Good communication is essential and works better when you know each other. This is particularly important when a large laboratory is used, as it is often possible to arrange to have work done by just one or two technicians and it is then easier to obtain an effective working relationship with these technicians. It is so easy to sound over-aggressive when talking to a stranger on the end of a telephone and differences of opinion can be solved much more easily when you know who is at the other end of the line.

To be able to organise the surgery appointments it is essential to know the laboratory arrangements. There are a number of questions which must be asked, either directly to the laboratory, or to the principal of the practice if a visit to the laboratory has not been arranged (Table 1). Some of these points will be covered in chapter 16.

The time of collection and delivery is important, so that appointments for working impressions, especially if for metal framed dentures, can be timed to coincide with collection and impressions are not left too long before casting. If there is no collection service and surgery personnel deliver and collect the work, then the appointments need to be arranged for the most convenient time, such as just before lunch or going home. Even if there is a collection service, if the practice can deliver without too many hardships, it may well be possible to trade this against the cost of free retries or something similar. If the practice has a large prosthetic workload, then there are several advantages to seeing all prosthetic patients on the same session: the DSA will be able to prepare the surgery for prosthetics for that session, collection and delivery are more cost efficient, and every one gets to know the

routine. When it is not possible to have a personal collection service, the only alternative is to use postal or courier services. These require further consideration, and will be looked at later.

To be able to book future appointments for patients, there must be a clear understanding of how quickly the laboratory will turn round the work. This needs to be expressed in working days, so as to take into consideration bank holidays and weekends.

If a small laboratory is being used and there is only one technician, it is unlikely that he will have cover during illness and this will obviously mean some rapid telephone calls to patients. Furthermore, if he takes his holiday at different times each year, it is essential to know this, so that patients are not promised in advance, treatment which requires technical support, such as a 'quick reline' to coincide with a patient's holiday. If the practice requires considerable technical support, the use of two laboratories may have advantages in offsetting some of these problems.

There are many variations in fee structures and how they are set, but they must be totally understood, especially if work

Table I Check list of questions regarding laboratory service

Is there a collection service and if so, what time and which days? If there is no collection service can you deliver and collect the work yourself?

What turn-around will they provide? Does this vary for different stages? If the work is being posted is the turn-around counted from their receipt of the work or from the time of your posting? Is any insurance being used and who is responsible to pay for remaking in the event of loss or damage?

What happens when the technician goes ill or on holiday? Does he take his holiday at the same time each year? If not, how much notification will he give?

What are the fees and what exactly do they cover? Will they charge for study casts and how quickly can they return them? Is a special tray included in the price? Which artificial teeth do they use routinely? How much is a retry or is it free?

What services do the lab supply? What articulators do they use routinely and what others are available? What equipment do they have? What condition is it in? Can they cope with advanced techniques if required? Do they subcontract any work?

Do they have a surveyor and do they use it? May you use it? Do they duplicate casts?

What is the standard of the work in progress on the bench while visiting?

is being carried out under an NHS agreement. The difference between a profit and loss under such a contract is extremely small, although of course the patient's treatment should not be tailored to a price.

So that time is not lost by work being sent to the laboratory for services which they do not supply, it is necessary to know about these in advance. It is also useful to know if they subcontract any of the work as it may be more reliable to send the work directly to the other laboratory or another of your choice.

The type of equipment which can be seen in the laboratory will give some indication of their usual services and the condition of the equipment an idea of the care they take. Even simple hinge articulators perform badly if the hinge is slack or worn. If there is no surveyor, or it is in a corner gathering dust and not being used, then fitting partial dentures can be stressful. There is nothing more frustrating or time consuming than trying to fit a partial denture which has been made on a cast which has been blocked out either incorrectly or not at all. Fitting a partial denture which has been finished on a duplicate cast and then fitted back onto the master cast saves considerable chairside time. It will also impress the patient when the denture goes straight in the first time, without adjustment. This may mean an enhanced technical fee, but this needs to be balanced against the savings in chairside time and raised blood pressure.

Using a postal system for casts is a dubious practice, even though impressions can survive and the dimensional changes which occur may not be clinically significant. If there is space in the practice to set aside an area for casting impressions, then it is quite possible to cast them immediately they have been taken. Very little equipment is required, although a vibrator and cast trimmer are useful. The casting can be done either by the practitioner or, with a little training, by the DSA. Even if a postal service is not used, such an area can be valuable for pouring study casts, which saves the technical cost of the study cast and means there is no time delay from sending the work out and receiving it back. An area for casting can also be very useful for making a quick repair for that 'special' patient. Packaging for posting can be a problem, but there are several ingenious custom-made systems which can protect casts through quite violent deliveries. These will be considered in chapter 16.

Postal services are not always reliable and, unless appointments are made with generous time intervals, it is sensible to ask the patient to telephone the practice to check that the work has been received. A simple system of logging work is suggested in the next chapter. Courier services do not have these disadvantages, but they are expensive.

Having selected one or more laboratories, it is essential to establish an effective and rapid means of communicating the prescription. It must be remembered that the practitioner is responsible for the quality and design of the prosthesis which is fitted. Although students are taught to prescribe, surveys have shown that at least for partial dentures very few actually do when qualified.[1] The reasons for this are many, but one of the main ones may be that the system used as an undergraduate was so standardised that its value was never appreciated, or because the ease of direct verbal communication with in-house technicians debased its value, or because the technicians were so experienced in the school's

Table II Suggested standard laboratory instructions

Casts
Pour all edentulous and immediate impressions in 50/50 mix and maintain all sulcus reflections (depth and width).
Pour partial impressions in stone.
Do not block out any undercuts unless requested.
Notify surgery of any teeth which are fractured when removing impression.

Special trays
Make all edentulous trays close fitting to full depth of sulcus unless marked. Make vertical stub handle and finger rests over the premolar regions.
For partials, use single wax spacer over edentulous regions with double spacer over the teeth. Extend to full depth of sulcus over edentulous areas and abutment teeth, but only to gingival margin elsewhere. Use a straight handle if anterior teeth are present, but a stub handle if only posterior teeth are present.

Record rims
Make with wax base and rim unless otherwise requested.
Lower rim should be centred over ridge and should be 20 mm high from the canine sulcus reflection.
Upper rim should be buccal to ridge and should be 22 mm high from the canine sulcus reflection.
In Kennedy 1 lower, add labial bow around anterior teeth (to prevent distal movement).

Try-in
Use average value articulator.
Use narrow mould posterior teeth.
Set upper incisal edges to edge of rim.
Set lower incisors with the necks over the ridge and the tips proclined to give stated over-jet.
Set lower premolars over the ridge.
For partials block out to given path of insertion.
Duplicate blocked out cast and use duplicate cast for all construction.

Finish
Post-dam as cut unless otherwise specified.
Wax down to full width and depth of sulcus unless otherwise shown.
Do not contour or stipple unless requested.
With partials, wax down and finish on or above survey line and fit back onto master cast.

Castings
Make rests with minimal thickness of 1 mm.
Lingual bar should be spaced 0·5 mm from cast.
All connectors should be rigid.

Immediate dentures
Pour casts in 50/50 mix and make special tray to full depth of sulcus.
Reduce cast as requested.
Maintain position and incubation of natural teeth to be replaced unless otherwise specified.

procedures that they produced the necessary work *despite* the prescription. On the other hand, it may be that in a busy practice it simply does not seem worthwhile, since no one else bothers. It requires discipline to take the trouble to write an adequate prescription.

Without a prescription, not only is it necessary to accept the design of the technician, who has rarely even seen the patient, but there is also no real contract, which makes it difficult to demand a free remake if the patient is unhappy about a feature which could have been designed differently. Of course, a discussion with the technician may often lead to a joint design, which encompasses the best features from the clinician and the technician. When a joint design is agreed over the telephone with a large commercial laboratory which receives many cases each day, one should not forget to write down the essence of the design, together with the name of the

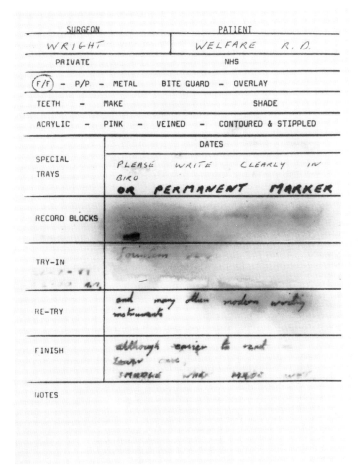

SURGEON		PATIENT		
WRIGHT		*WELFARE*	*R. D.*	
PRIVATE		NHS		
(F/F) - P/P - METAL	BITE GUARD - OVERLAY			
TEETH - MAKE *T N R*		SHADE *O 3*		
ACRYLIC - ~~PINK~~ - VEINED - CONTOURED & STIPPLED				
	DATES			
SPECIAL TRAYS	*PLEASE WRITE CLEARLY IN BIRO* **OR PERMANENT MARKER**			
RECORD BLOCKS	*Others such as water based ink markers or*			
TRY-IN 21-7-89 12.00 A.M.	*Fountain pens or roller balls & letter pens*			
RE-TRY	*and many other modern writing instruments*			
FINISH	*although easier to read in lower case,* SMUDGE WHEN MADE WET.			
NOTES				

Fig. 1 Specimen prescription form—text written with various examples of modern pens and 'biro'.

SURGEON		PATIENT		
WRIGHT		*WELFARE*	*R. D.*	
PRIVATE		NHS		
(F/F) - P/P - METAL	BITE GUARD - OVERLAY			
TEETH - MAKE		SHADE		
ACRYLIC - PINK - VEINED - CONTOURED & STIPPLED				
	DATES			
SPECIAL TRAYS	*PLEASE WRITE CLEARLY IN BIRO* **OR PERMANENT MARKER**			
RECORD BLOCKS				
TRY-IN				
RE-TRY	*and many other modern writing instruments*			
FINISH	*although easier to read lower case,* SMUDGE WHEN MADE WET			
NOTES				

Fig. 2 Same form after 24 hours in damp denture bag. 'Biro' is still completely legible, some water based inks have totally washed out, permanent markers have spread slightly.

technician with whom it was arranged, as well as the time and date of the telephone call. There is no point in sending work which will arrive 2 or 3 days later along with hundreds of other cases, with 'as discussed' written on it.

If the same laboratory or technician is being used all the time, it is possible to have some general standard instructions which can be given at the beginning. This can save considerable time and helps to define a contract with the laboratory (Table II). It is worth producing these in print, and even sealing them in a polythene bag, so that the technician can, if necessary, refer to them with wet hands.

Just in case there is ever a query, remember to keep a copy at the surgery for reference and updating. Most laboratories produce their own forms to help with their administration and to encourage the practitioner to write down the basic instructions. Some forms are quite complex and require time just to work round all the boxes. Nevertheless, these should be completed. In most instances, the simpler the form is, the better.

Even experienced clinicians will forget to include some detail if they do not have a routine. Routine is essential and the DSA should be instructed to fill in the name of the surgeon and patient, the service required, either private or NHS, the stage of work, and the date and time of the next appointment. Some laboratories like the date to be the day before the patient's appointment, so it must be known whether they are working with their clock 24 hours fast so they are not late! If the DSA is ill and A. N. Other fills in the form, if there is a routine, then, again, information will not be forgotten. While considering basic details, it is worth remembering that the prescription, to be of value, must be legible. Many of the modern 'pens', although they produce a pleasing appearance on dry paper, have water soluble inks which, when damp, have a tendency to smudge badly and may become illegible or even totally 'washed out' (figs 1 and 2). Most laboratory work becomes moist at some time and it is vital that the prescription does not, despite remaining attached to it. Old-fashioned 'biros' are probably best for completing the prescription, although permanent markers can be used. Poor handwriting is another liability, and the use of capitals is advised. Figures 3 and 4 show two prescriptions which were completed as part of a survey by Walter[2] on partial denture design. The prescription in figure 3 was universally disliked by a team of technicians when asked to fulfil the request, whereas the one in figure 4 was universally liked.

Figure 1, as well as illustrating the value of the simple biro, also shows a simple prescription form which is quite adequate for most cases and any other instructions not included on the standard prescription can easily be entered at each appointment.

The prescription
In addition to the standard instructions, certain points may need to be inserted or modified.

Complete dentures
Special trays which are required to a specific outline can be designed on the impression. It is quite easy to draw on the impression with an indelible pencil or permanent marker. These lines will then transfer to the cast when poured. It is best

to dry alginate impressions by blowing, otherwise they are difficult to mark. Water soluble pens mark easily, but run as soon as they are made wet and give a very broad line on the impression and an even broader line on the cast. Some marks tend to fade quickly after the cast is poured, so the technician needs to be advised that the impression has been marked so that he can reinforce the mark on the cast soon after it is poured.

If the artificial teeth are ordered together with the special tray or record rim, a retry may be avoided because the patient does not like the shade or mould of the selected teeth. If the patient is unhappy, not only can the teeth be returned without cost, as they have not been ground, but also there is more chance that the second choice will be to the patient's approval.

Record rims require little more than that provided with the standard information, but remember that the more rim there is, the longer it takes to trim.[3] If a denture is to be made with a large free-way space, then the technician should be asked to make the rims shorter. If the patient has an obvious Angle's Class III skeletal relation with a large mandible, the upper rim should be asked to be set as far buccal as possible, in order to make the subsequent shaping and recording procedure a little easier.

If the patient has always had problems with the retention of a denture, it is sensible to have that rim made on a heat-cured base so that retention and stability can be checked early. If retention or stability are still a problem, then it is not worth proceeding until the problem has been solved, as it will certainly not improve just by finishing the denture. The heat-cured base will also make the jaw registration much easier and less frustrating, as it will not move during the recording. If a heat-cured base is used, the final processing must be a long slow cure and not one of the quick cure methods, otherwise the base may warp. Alternatively, if there is limited inter-occlusal space, wax bases, made by repeatedly dipping the wet cast into molten wax, may be used.

Try-in details must include the shade of the teeth and, ideally, the mould also. However, selecting the mould produces problems at the beginning of practice life, since very few surgeries keep a mould guide because of the cost. Fresh from dental school and used to the luxury of complete mould guides, the new practitioner is left struggling with inadequate paper mould guides. There is no quick solution to this problem, but it is essential to decide on the make of teeth which will routinely be used, preferably one stocked by the laboratory, and then learn to picture the shape and size of the most popular teeth on the guide, so the mould can be selected more easily. It is also worth collecting any sets of teeth which have been ground for a try-in but have then had to be changed because the patient did not like them. These can be used to make a personal mould guide.

Mould selection can be delegated to the technician if the patient is happy with the old dentures and an impression is taken of it. This method will often incur an enhanced fee for casting the impression and may result in copying something which is not ideal. The information on shade and mould should also be entered on the patient's notes for future reference, in case the patient ever requires a new or a spare set of dentures. If the upper rim is trimmed to give the correct length and position of anterior teeth, the only extra information the technician needs to set up these is the overjet and over-bite. This will be governed by the degree of

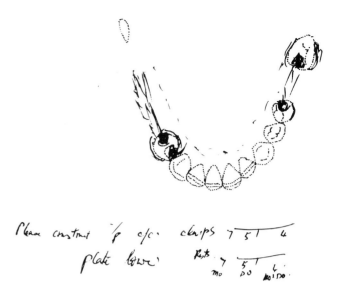

Fig. 3 Prescription form universally disliked by team of technicians trying to complete instructions.

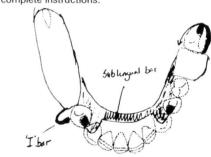

Fig. 4 Prescription form universally liked by team of technicians trying to complete instructions.

resorption of the ridge, the desired appearance, the lower lip activity and the patient's denture control. Generally, a flat ridge will require less overbite, and poor control and greater lip activity, greater overjet. The more the patient wants a natural appearance, the more these rules will have to be compromised and the patient warned of the limitations of this.

With regard to the support given to the cheeks by the record rim, the position should be determined in the mouth and the technician should understand that the shape so produced is the prescription. For example, 'set the teeth further buccally' is inadequate.

If a special articulator is required, the technician will not know this unless it is requested. One disadvantage of not having a technician on the premises is that try-ins are returned off the articulator. This may present problems if teeth need to be moved. Some articulators, although more expensive than a plane line articulator, have removable mounting plates, making it possible to keep one articulator at the laboratory and one at the surgery, provided that they are standardised in the first instance.

Finish instructions require the position of the post dam to be shown, although few dentists bother to do so.[4] Ideally, the post dam should be cut on the cast while the patient is still in the surgery. In the case of a rebase, it will have to be delegated to the technician and then precise instructions must be given and the cast clearly marked. If the laboratory does not routinely cure their dentures overnight and a slow cure is required, then it must be asked for. Contouring and stippling should only be requested when the patient is likely to show considerable amounts of denture base, as it soon becomes stained if the patient's denture hygiene is not perfect. The denture should be finished to the full depth and width of the recorded sulcus, but in those cases where an overextended impression has been accepted, or where a conscious decision to make the denture flange short has been made, its required position should be clearly drawn on the cast.

Partial dentures

Apart from the obvious differences between the instructions for complete and partial dentures, the important extra information required by the technician is the path of insertion and denture design.

The path of insertion requires the cast to be surveyed and marked on the side to identify that selected. This needs two things: first, a cast and, secondly, a surveyor. The standard teaching at dental school is to have study casts made and then to survey and design the denture.

If it is necessary to have a special tray constructed, the easiest method is to use the primary cast, providing the technician is asked to remove the wax spacer and not to damage the cast while making the special tray. There are, of course, times when special trays are not needed, especially if the denture is a tooth-supported metal skeleton denture, in which case stock trays are often perfectly adequate. In cases where a replacement denture of a similar design is to be made, and a wax try-in for tooth position is needed, the working cast can be surveyed at that time, since no tooth modification will be needed In other cases it is essential to take an alginate impression and make study casts. It is particularly useful if these can be produced at the surgery, as they can be quickly cast up and analysed before the patient's next appointment and the treatment plan prepared in advance rather than doing this hurriedly while the patient is in the chair. Then, if any tooth modifications are required, to allow for rests or to create guide planes, the surgery can be prepared for this extra treatment. To analyse the casts it is necessary to have a surveyor. These are not always cheap, but inexpensive surveyors can be bought from £179. This cost has to be set against partial denture fees and, where returns are limited, such as on NHS contracts, there is often a reluctance to purchase a surveyor. If the laboratory is close to the surgery, it may be possible to use their surveyor, but this is inconvenient unless visits are being made by the dental surgeon to take work regularly to the laboratory. Failure to use a surveyor means that a denture cannot be properly designed and the necessary tooth preparation for guide planes and more effective clasping cannot be planned. Handing the

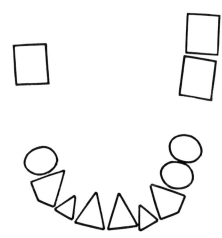

Fig. 5 Simple diagrammatic representation of partially edentulous mouth on which to draw denture design. Teeth are shown as different basic shapes.

responsibility to the technician means that the design will not allow for the clinical condition of the teeth. Also, if there is a problem with the design, the dental surgeon has little room for complaint.

The design for the denture is best drawn on the study cast, although if there is only a master cast it is advisable not to mark this cast but to draw the design on a sheet of paper. If a proper design sheet is not available, teeth may be simply depicted: molars as oblongs, premolars as ovals, canines as diamonds and incisors as triangles. Rotation of teeth is shown by rotating the particular shape accordingly (fig. 5).

With regard to the prescription, it must be remembered that the technician is also busy and the prescription should be as brief as possible. Instructions should be written in the order the technician will carry out the work, so it is not necessary for him to read the whole prescription each time he refers to it. Also, as he carries out each instruction, he can cross off each completed request.

One very difficult aspect of prosthetic dentistry is quality control of the laboratory and clinical work. The present NHS fee scale forces both the technician and practitioner to cut costs to the limit and there can be a tendency to omit some stages of denture construction in an attempt to save time and money. Experience helps to predict when this is safe, although it is not infallible. Accepting poor standards at the start of practice life soon makes the attainment of high standards impossible. Practitioners, once they have accepted poor impressions or delegated design to the technician, will soon forget the fundamentals of their undergraduate training. Also, if substandard technical work is accepted and used in the clinic because returning the work will mean another appointment and further cost, or the clinician does not have the courage to say 'No, this does not meet my standards', then the laboratory is unlikely to improve their own quality control. As soon as work arrives at the surgery from the laboratory, it should be established that it is the correct piece and that the necessary work has been done. The work should then be checked to see that it has been completed to a reasonable standard, while there is still time to postpone the patient's appointment if necessary. Work which fails to achieve an acceptable standard should then be brought to the attention of the technician. Otherwise, frustration and disillusion sets in, as standards are not met and satisfaction with prosthetic dentistry is lost. Under the present system, if experience is to be gained without stress, then it must be understood that on some occasions, when constructing dentures under an NHS agreement, there will be a financial loss. With greater experience, safe short cuts will be learnt and difficult patients recognised.

Good clinical work cannot be achieved without good technical support, nor can good technical work be provided with poor clinical support. Skimping on technicians' services or fees is likely, in the end, to take more clinical time and cause irritation to both practitioner and patient.

References

1 Barsby M J, Schwartz W D. A survey of the practice of partial denture prosthetics in the United Kingdom. *J Dent* 1980; **8:** 95-101.
2 Walter J. Personal communication, 1989.
3 Halperin A R, Graser G N, Rogoff G S, Plekavich E J. *Mastering the art of complete dentures.* pp94-96. Chicago, Illinois: Quintessence Publishing, 1988.
4 Murphy W M, Bates J F, Huggett R. Complete denture construction in general dental practice. *Br Dent J* 1971; **130:** 514-521.

16

Prosthetic Dentistry: In the Surgery

R. D. Welfare and S. M. Wright

It is important for the smooth running of appointments that everyone, patient, receptionist, dental surgery assistant (DSA), the laboratory and, of course, the dental surgeon should know what is happening. Obviously, they do not all require the same information, but certain points will be essential to each of them for maximum efficiency of treatment.

When a patient makes an appointment for dentures, whether this is by telephone or in person, his first contact is with the receptionist. It is important, therefore, that the receptionist should have been given a note of the probable minimum number of appointments needed for complete and partial dentures and a guide as to the likely cost of such treatment, so that she will be able to inform the patient. Before making an appointment for a new patient, it is helpful if the receptionist asks why the patient wishes to see the dentist. For example, if the patient wants new dentures before his holiday in a fortnight, it is better to say immediately if it is unlikely that this can be achieved. It may be appropriate in the case of a patient requiring dentures for an appointment to be arranged for a particular session, in order to fit in with the laboratory's collection days or times. It can also be useful to know whether the patient has any of his own teeth remaining, since these may require treatment before denture impressions can be made. If, for example, different lengths of appointment are used for private patients, the patient may be asked whether he/she wishes to have private treatment or to be treated under the National Health Service. New patients should be asked to come a few minutes before the time of their appointment and to bring their reading glasses, if they need them, in order to complete a medical history questionnaire.

On the day list, the receptionist should indicate the patient's likely requirements, such as complete dentures, partial dentures, and so on. This allows the DSA to have the necessary materials and equipment to hand. To assist the DSA in this task, especially if she is also new to the work, the dental practitioner should provide her with a list of his routine needs for each clinical stage. In itself, this is a good exercise as, if domiciliary treatment is undertaken, it is important that only essential items are carried and that none of these are forgotten.

Good communication between the dental surgeon and the patient is essential if misunderstandings are to be avoided. What are the patient's expectations from denture treatment? It is important that the dentist should appear interested in the patient's wants. He/she may show this to the patient by verbal tone, and general demeanour, which together make up body language. Body language can be used to encourage a patient to express his/her problems and desires with regard to dental treatment. It is also the responsibility of the dental surgeon to read the patient's body language and to use his/her training to be objective about the patient's problems. Is an aggressive patient really terrified, or a talkative patient in fact very depressed?

Patients are likely to be very subjective about their dental problems. After all, it is *their* mouth, and they know no other,

even if they have seen and heard of other people's apparently successful dentures. Can patients' expectations be fully realised? If they are expecting miracles and the reality of the situation is not explained so that they understand it, disappointment is inevitable. It is important that the practitioner appreciates patients' feelings, concerns and worries and considers these in his explanation of what may or may not be achieved.[1] It is a good policy for the dental surgeon to tell patients at the examination what can be achieved and what they can reasonably expect. For example, if a patient's appearance can be improved, but the stability of the dentures cannot, he/she should be told and this must be entered into the notes. The patient is then less likely to blame either the dentures or the dentist for failure.

This communication is an extremely important part of the initial history and examination phase of the treatment. It lays a foundation of trust between the patient and the dental surgeon, on which successful treatment can be built. The kindness and consideration which the dentist shows at this stage takes little time, yet considerably helps the patient to relax, and so facilitates easier recording of impressions and, later, of the centric jaw relationship. Unfortunately, misunderstandings about denture treatment may result in conflict and even in litigation. These are especially common in the provision of immediate replacement dentures. The patient may not have fully comprehended the fact that, normally, such dentures become looser as resorption occurs, and that relining may be needed. He/she may not understand that in order for the dentures to be relined, they will have to be taken away and sent to the laboratory for reprocessing, and that an additional fee will be charged for this service. It is important that the patient should be educated to realise that the dentures are temporary and will need to be remade at some future date, and periodically after that. The Medical Defence Union[2] has advised a practitioner to consider asking patients being fitted with immediate dentures to sign a simple consent form as evidence that a proper explanation has been given (fig. 1). Alternatively, it would seem sensible to give them a written information sheet at their first visit. Written information can also be usefully given to other denture patients.

With my agreement Mr.......................... is inserting dentures immediately after extractions. I understand that as a result of further slow shrinkage of the gums, the dentures now being supplied may become ill-fitting and may have to be relined or replaced at additional expense to me.

Signed..

Date...

Fig. 1 Form of consent recommended by the Medical Defence Union.[2]

At the first visit, the dental surgeon should make brief notes of relevant details of the patient's history, unusual features of the mouth, and any other points which may affect the treatment. The treatment plan should be recorded simply, so that it may be referred to quickly during the course of treatment. For example, in complete dentures, intended changes in base extension, vertical dimension or tooth position would be noted. Obviously, any treatment carried out during that visit should also be recorded. It is helpful if the planned appointments, their minimum separation and the duration of each is also listed. This may be written on the notes or by the DSA on a suitable surgery-to-reception memo, which the patient will take to the receptionist to book the required clinical appointments and also to make any necessary laboratory booking, for example, for a metal casting.[3] It is helpful if the patient understands that if an appointment is later cancelled, this may affect the timing of subsequent appointments. For example, if the laboratory requires 2 weeks to make a casting and all other appointments are weekly, and an appointment is cancelled before the date for the working impression, some of the following dates will need to be changed and the casting rebooked with the laboratory.

It is also important that the patient is told the amount of payment that will be charged and when it is due to be paid. For larger amounts, or for private treatment, a written estimate is helpful, which should include expectations and limitations of treatment. For NHS patients, the EC17 form must be completed.

The mixing of National Health Service and private treatment is to be avoided as it is a perennial source of difficulty. As a general rule, if a patient specifically requests certain treatment under private contract and some within the National Health Service, the private work should be completed and paid for before the patient is accepted for treatment within the NHS. The dental surgeon should realise that once the patient has signed the NHS form, he/she has a responsibility to make the patient dentally fit, even if the patient is a difficult one!

The matter of remuneration under the National Health Service is emotive and fixed. There is no such thing as an 'NHS denture'.[4] The dental surgeon's reputation is at stake and he/she alone is responsible for the quality of the prostheses placed in a patient's mouth. It may be expedient to remember that under the Sale of Goods Act, which covers dentures, if the dentures are not fit for their purpose, the patient is entitled to his/her money back. In addition, products which are not considered good value for money can give rise to problems relating to fees. A pleased patient will tell others and good prosthetics is practice building.

Practice management

If, at the first visit, impressions are taken, a laboratory form must be properly completed. Items which are to be sent to the laboratory should be labelled by the DSA with the dentist's and patient's names and should be clean and neat. Obviously, the standard of work received by the laboratory is likely to affect the technician's perception of the standard of work the practitioner expects to be returned. Impressions should be carefully rinsed to remove all traces of blood and saliva. It is unfair to expect a technician to receive and cast up an impression with blood at the gingival margins and thick mucus on the palate.

The impressions should be sealed in a polythene bag with the label stapled to the *outside* of the bag, so that it does not become wet and illegible. Similarly, record blocks should be free of loose pieces of wax and the record should be clear and stable. If the work is to be posted to the laboratory, it is recommended that non-stable impressions are poured (even if they are not based) at the practice. It is useful if the DSA is trained to do this. There is often time for her to pour the impressions for one patient while the practitioner is recording the jaw relationship for the next one. All work should be suitably packed for safe transit. For delivery to, or collection by, a local laboratory, the packing can be simple, for example, the bases of two one-pint plastic milk bottles may be used. Four of these fit neatly into an oblong 2-litre ice cream box and each will hold, separately, the work of one patient. If the work from several patients is heaped together in one box, instructions easily become separated from the related work. Equally, where items are crammed tightly together, they may become distorted or damaged.

A note should be made of the date that the work is sent to the laboratory, together with details of the work that is required and the date when it should be returned. This may be written in the day-book or in a special book kept for quick reference in the case of a query. If the work is to be posted,

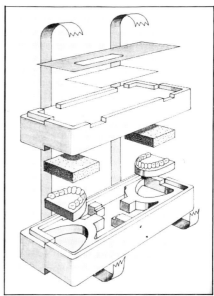

Fig. 2 Quick-pak lightweight packaging is suitable for posting denture work.

purpose-built, light-weight packaging* is convenient (fig. 2). Many laboratories provide address labels with free prepaid first class postage. The Post Office will issue a free Certificate of Posting, which provides evidence of posting and shows the date of posting, but compensation will not exceed the market value of lost items to a maximum limit of £20.

For items worth more than their material value, Registered Post with Consequential Loss Insurance will provide cover of up to £10 000.[5] Where a faster service is required, an extra fee may be paid for the Royal Mail Special Delivery[5] or, for an even larger payment, a courier service which provides insurance may be used. Obtaining recompense for the wasted

---

* Details can be found at the end of the chapter.

surgery time and effort might be difficult, even with insurance.

Each afternoon, the dental surgery assistant should check the following day's list and ensure that the laboratory work is available for every patient for whom it will be required. Checking in this way means that there is still time to cancel a patient and avoid his making a wasted journey if, for some reason, the work is not ready. Those patients for whom a postal laboratory service is being used should be asked to telephone on the previous afternoon to confirm that their work has arrived (again to avoid a wasted journey).

Cross-infection control
The control of cross-infection in prosthetic dentistry may be difficult, as much of the equipment was designed before today's knowledge of viral diseases. By providing the DSA with a list of the most usual needs for different procedures, the practitioner can reduce the amount of equipment that will need sterilising or disinfecting after each patient's treatment. The use of the zone system to reduce cross-infection risk is helpful.[6] Covering the work area with a sheet of paper which is disposed of after treatment means that wax drips and debris are simply removed. The surface should then be cleaned and dried with a solution containing 70% isopropyl alcohol.[7]

Obviously, it is impractical to sterilise or disinfect some items, such as the patient's notes. These should be handled by the dentist *before* he touches the patient, or afterwards when he has washed his hands again. Impressions should be taken in disposable trays. They, like the appliances, should be rinsed thoroughly to remove all visible blood debris before sending to the laboratory. Technicians (and DSAs) should be encouraged to wear gloves when handling impressions and pouring models.[7]

In addition, it is recommended that work is sprayed with an antiseptic spray prior to it return to the technician. It may also be worthwhile spraying incoming work. A chlorhexidine isopropyl alcohol based spray may be used.[6] The use of specific individual containers to keep each patient's work separate is also helpful.

Quality and efficiency
It is extremely difficult for any dental surgeon to carry out good prosthetic treatment when he feels tense and rushed. Somehow his tension seems to communicate itself to the patient, making it harder for the patient to relax. The patient's tension will then affect the dentist and a vicious circle is soon under way (fig. 3).

While experience will allow some short cuts in procedure, it is probably better to begin by accepting lower returns and to concentrate on trying to assess objectively the quality of each piece of work as rapidly as possible. Some patients seem to be able to wear dentures despite the efforts of the dentist, but equally there are patients who can only become successful denture wearers when provided with correctly constructed and properly functional dentures. The majority of patients lie somewhere between these two extremes. Since there is no simple way of distinguishing 'good' from 'bad' denture wearers, it must be assumed that all patients fall into the 'bad denture wearer' category and always to aim for a good standard of treatment.

Successful impressions are only possible in one visit if the

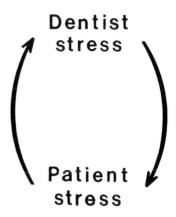

Fig. 3 Stress—the vicious circle.

dental surgeon is able to recognise (and then to achieve) correct extension in the distolingual, buccal, anterolabial, tuberosity and post-dam regions. A large selection of trays can help in accommodating mouths of unusual form and dimension. Plastic disposable trays can be cut, or modified by tracing their borders to make them fit the patient better, before the impression is attempted. If a proper impression cannot be achieved in a stock tray, then either the use of a two stage impression (for example, impression compound trimmed back and covered with an alginate wash) or of a special tray must be considered. Alternatively, in edentulous cases, impressions may be taken in the patient's own dentures. This is possible only when the dentures have no undercut to prevent their being removed from the cast. To be used as impression trays, a patient's dentures must be properly extended or have borders that can easily be built up before the impressions are taken. Obviously, the patient will be without his dentures while these impressions are being poured. They may be poured in the surgery while the patient waits in private if he is without his teeth. Sometimes, the patient may have an older set of dentures which can either be used for the impressions or which he can wear while the impressions are sent to the laboratory for casting.

Where copy dentures are planned, again the DSA can be trained to assist in making the moulds and even in pouring the duplicates in acrylic. However, if a technique is employed which uses alginate impression material in a box, once the denture is removed the box may be closed, sealed in a labelled polythene bag with a moist tissue and sent for pouring to the laboratory, together with the lab card.[8,9] Obviously, if a copy technique is to be used, the practitioner must be certain which aspects of the denture he wishes to copy accurately and make a considered decision as to which areas require active modification. In addition, he must clearly specify his needs to the laboratory. A copy technique is not necessarily an easy way out of making dentures, since it requires careful attention to detail, in both the clinic and the laboratory.

Where impressions are taken for partial dentures, it is essential to use fixative, so that the impression material remains fully attached to the tray. Where study casts are required to decide on tooth modifications, it is important that the laboratory knows that orthodontic trimming is not needed, to avoid an extra charge. Alginate working impressions for metal dentures should be cast immediately to ensure accuracy. The clinical time taken to record the centric jaw relationship will be reduced if the rims of the record blocks are made to reasonable dimensions in the first instance (see previous chapter).

The old dentures may be used as a guide to the rim heights (with due consideration for any planned alterations). Unnecessary bulk, especially at the distal end of the rims, should be removed, as it will cramp the tongue and make it more difficult to record the correct retruded position. Removal of the distal part of the rim also avoids displacement of the lower base because of pressure on the distal slope of the mandible when the record is made (figs 4 and 5). The relation of the rims to the lips and cheeks and the amount of support provided should be noted and the prescription adjusted accordingly. If the impressions have been recorded in the patient's own dentures, a check record may be made at that time. The dentures on the casts can then be articulated, provided the patient is prepared to be without his dentures while this is being done. If the tooth position is to be copied, it must be ensured that their relationship to the ridge has not been altered by an uneven thickness of impression material.

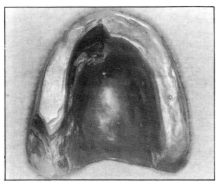

Fig. 4 Modification of the upper record rim gives more tongue space (*left*).

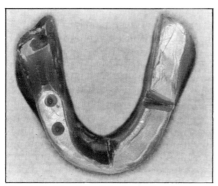

Fig. 5 Modification of the lower rim gives more tongue space and avoids displacement of the lower base.

When the mouth of the partially edentulous patient is examined, prior to the recording of impressions, it will be seen whether the occlusion is stable and there are enough suitably sited remaining teeth to allow the cast to be articulated without the use of record blocks, or whether only one record block will be sufficient to provide a positive recording. Articulation of the casts can only be as accurate as the record that is sent to the technician. Sometimes, problems are anticipated with the stability of the record blocks, for example where the ridges are resorbed and there is excessive activity of the tongue and circumoral musculature. If the patient is partially dentate, with bilateral free-end saddles, a wire placed buccally to the remaining teeth to connect the two

saddles may be incorporated in a conventional record block to prevent distal movement and bending of the record block. For an edentulous patient, a heat-cured acrylic base may be used to prevent distortion. A 'wax dip' technique also provides a well fitting base which gives improved stability.

At the try-in, all parameters of the dentures should be checked. Patients should have been asked to bring their glasses if needed to see the try-in. They should be given adequate time to look at the appearance of the dentures. If they are doubtful, or need more time, it is better to let them take the try-in home, rather than to rush to have the denture processed. It is important that the patient is happy with the appearance, feel and comfort at the try-in stage, and their comments should be considered. 'The teeth meet too soon' may indicate an excessive occlusal vertical dimension. It is very much cheaper and easier to repeat a wax try-in than to remake an acrylic denture. Finally, the post dam should be cut into the cast with reference to the patient's tissue displaceability, and the denture sent for finishing.

Before the dentures are inserted, the fitting surface should be checked. Any sharp prominences, including the patient's name if this has been carved into the model, should be removed to prevent trauma.[10] If the patient feels any areas of discomfort, these should be identified using a pressure indicator paste, with no tooth contact, and eased. The occlusion should then be re-examined. The teeth should meet evenly. This is easily checked by asking the patient to close but then to stop when the upper and lower front teeth are still separated by 1-2 mm. This will accentuate any uneven contact (figs 6 and 7). If the teeth do not meet evenly, remounting to

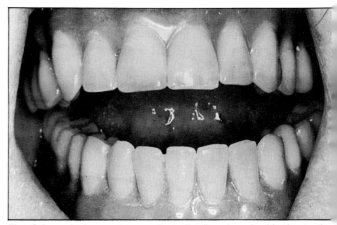

Fig. 6 A marked premature contact distally, viewed with the teeth slightly separated.

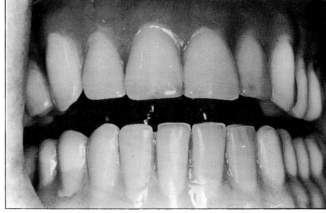

Fig. 7 An even occlusion viewed with the teeth slightly separated.

a precentric check record and grinding it to an even centric occlusion is advisable, as this removes the likelihood of an increasingly dissatisfied patient re-attending with escalating regularity, to complain to all those in the waiting room, as well as to the dental surgeon, about his painful dentures.

At the 'fit' appointment, the occlusion should also be checked for freedom in lateral excursion and the absence of locking on the canine teeth. The patient should be given advice on the care of new dentures. If there is pain, he/she should be asked to wear the new dentures during the daytime of the 24 hours prior to the review appointment, so that any sore areas may be more easily identified, and to bring the old dentures. A written sheet giving this information may be useful.

At the review appointment, the patient's comments may once more be very helpful in locating the cause of a problem. For example, a 'sore throat' may indicate over extension distally of either upper or lower dentures. Apart from checking the mouth for inflammation and ulceration, the dentures should also be checked. Small shiny facets on the teeth may give evidence of interferences which limit lateral excursion and should be adjusted. As at the 'fit' appointment, the quickest and most effective way of overcoming errors in the centric jaw relationship is to make a new registration and return the dentures to the laboratory for a grind-in.

Rebasing of newly fitted dentures is rarely necessary, unless the cast is poor or has been damaged. A denture is much more likely to be loose because of an occlusal error or because the base is over- or under-extended or there is an inadequate post dam. If the border is under-extended, then modifying it by tracing will resolve the problem. When the denture is fitted, the length of time booked for the review appointment may be adjusted according to the dental surgeon's anticipation of the patient's likely problems. The review appointment is a good practice builder, since the patient feels he/she is being cared for. It is also very educational for the dental surgeon, since he/she can see how successful the treatment has been. It is important to deal with as many problems as possible *before* the dentures are actually fitted and given to the patient. Patients should have been prewarned of any likely difficulties, so that, later, they do not feel that they are merely being given a series of excuses in answer to their complaints. The review appointment allows adjustments to be made to the dentures. Practitioners who are vocational trainees are fortunate since, where they expect difficulties, they may call on their trainer's experience to confirm their own decisions as they progress. While difficult patients cannot be treated profitably under the NHS, until the dental surgeon gains experience and builds a good reputation, he/she is unlikely to attract many private patients, other than those who are not very discriminating.

References

1 Grace M. The importance of good communication. Medical Protection Society Annual Report and Accounts 1987; **95:** 49-52.
2 Medical Defence Union 1982: 66-67.
3 See chapter 17.
4 *BDA News.* March 2nd, 1976.
5 Letter Rates. A Comprehensive Guide UK. Royal Mail 1990; PL(B) **4272:** 5-6.
6 See chapter 8.
7 The control of cross infection in dentistry. *Br Dent J* 1988; **165:** 353-354.
8 Heath J R, Basker R M. The dimensional variability of duplicate dentures produced in an alginate investment *Br Dent J* 1978; **144:** 111-114.
9 Davenport J C, Heath J R. The copy denture technique. *Br Dent J* 1983; **155:** 155-162.
10 Cavalier M. Early neoplastic change related to a denture identification mark. *Br Dent J* 1976; **140:** 23-24.

Quickpak: 35 Nettles Lane, Frankwell, Shrewsbury SY3 8RJ.
Dispray: Stuart Pharmaceuticals Ltd, Alderley Road, Wilmslow, Cheshire.

17

Management Science

M. D. Wilkinson

Many dentists believe that they are poor businessmen. Some even take the view that business is none of their business. Neither attitude is realistic. This is the first of six chapters which, by outlining some basic ideas, will demonstrate that efficient practice management is not only desirable, but also achievable.

Dentists are not alone in believing that their training for management is inadequate; most businessmen will admit that they too have had no formal training. However, although there is always scope for developing business skills, we already possess considerable ability. There is no mystery to the considerable organisational talent required to arrange a holiday abroad, to enter into a practice relationship, or to buy and set up a home.

In addition to clinical responsibilities, the general practitioner is required to run the business—organising staff, responding to patients' demands, planning treatment and arranging for its provision—and to cope with many problems, crises and even success.

Management is the art of organising people and events in a structured way. First, by identifying what you want to achieve. Second, by working out ways of attaining those objectives and, finally, by maintaining control and guiding, reviewing and adjusting when necessary.

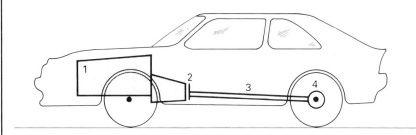

THE DENTIST AS THE MOTIVE POWER
Does Input Equal Output?

1 Engine: The dentist (input)   **3 Drive shaft: Delegation**
2 Clutch: Administration   **4 Final drive: Result output**

The dentist and his staff are analogous to the engine and transmission of a car, respectively. In the ideal car, the power output of the rear wheels is exactly the same as the power input of the engine, but in reality, there is a diminution in output which is directly proportional to the inefficiencies to be found within the transmission. Like the automotive engineer, the dentist should strive to maximise the efficiency of the transmission.

These two waiting rooms are very different but both give an impression of efficiency.

Why manage?

A well organised environment allows dentists and staff to produce their best work. Patients are impressed by the competence of the staff and the efficiency of the practice. Less time, materials and money are wasted. The practice team is well motivated and job satisfaction is reflected in a low turnover of staff, while patient satisfaction is exemplified by a high level of recommendations. Although the two practices below are very different, they both convey the impression that they are well managed.

Good management leads to greater efficiency, and efficiency is essential in dentistry, where only dentists can create income and hygienists represent the only opportunity for productive delegation. This dependence upon the dentist is responsible for much of the considerable stress within the profession.

One of the characteristics of a poor manager is an inability to delegate, because he or she cannot trust others. To work as productively as possible, dentists must delegate those tasks inappropriate to their skills. The dentist who types his own letters or completes his own forms is an over-paid clerk, while the person recruited to perform that task becomes demoralised through boredom and lack of responsibility.

Administrative routines

Every procedure in every walk of life can benefit from the establishment of routines to ensure that tasks are performed with consistent efficiency. On the other hand, routines can become so monotonous that initiative and operator satisfaction are impaired. The maintenance of this delicate balance is an essential managerial responsibility.

Administration can be defined as the instrument of management. Good administration consists of MRIR:
- Method
- Routine
- Instructions
- Review

Government administration can be unwieldy, inappropriate, slow, inefficient and unresponsive. This has lead to an association of administration with inefficiency. 'Bureaucracy' has become a derogatory term. 'No administration is good administration' has become a watchword. If MRIR principles are followed, this need not be so.

Method

Before any procedure is initiated, considerable time must be given to thought, discussion and trial. What is it designed to achieve? Are there precedents? Are there other approaches?

Routine

To minimise the opportunity for error, method should be translated into a routine. As an operative becomes able to perform the task automatically, he or she is freed from the worry that something has been omitted and, because a routine has become established, the task is performed more quickly.

Instructions

Written instructions are essential to establishing a correct routine. Staff cannot be expected to perform their duties to the dentist's satisfaction unless they have been given instructions. But the day is pressured, the patients are anxious and staff are required to be all things to all people in all places at all times. Moments when dentist and staff members are free simultaneously are rare.

Written instructions solve these difficulties. They can be prepared carefully, away from pressures. They can be examined and questioned, read and followed by any person at any time and they are an invaluable aid to staff training. The time taken to prepare written instructions may seem a burden, but they save much more time and avoid misunderstandings in the long run.

Review

Review should consider whether the procedure is still:
- necessary
- appropriate
- easy to operate
- being carried out.

Perhaps the most significant characteristic of poor administration is the failure to monitor its systems. 'Why can't they do something about it? Surely they can see it's stupid?' is a frequent complaint. Even dentists set up services which are not immune from such complaints. Circumstances change—replacement staff may bring different skills to bear, the need for a particular routine may disappear.

The confidence and self-esteem of staff will be enhanced if they are encouraged to monitor and discuss their routines. They will always respond to being consulted and, even when the dentist's wishes overrule those of the staff, that ruling will be better understood and accepted if accompanied by consultation and explanation.

Practice manuals

The logical development from a need to issue written instructions is the compilation of a manual. The advantages are considerable. They ensure uniformity of routines, simplify staff training, are an *aide-memoire* for permanent staff and an instruction manual for temporary staff, and provide a baseline for review.

The best type of book has a tough vinyl cover with transparent vinyl A4 pocket pages welded to it, as opposed to loose-leaf ring binders from which the pages can be removed and lost. By typing or writing on only one side of the paper and inserting two pages back to back in each vinyl pocket, it can be easily re-arranged or re-written at any time. Table I offers a list of suggested subjects for such a manual. Table II shows an example of a receptionist's duties.

Table I Suggested topics for a practice manual

Abbreviations eg DTC, FTA, OAF, FPC, EC17, MOD, PenV
Reception duties
Patient telephone routine
Laboratories eg laboratory tickets, collection times
Surgery/reception instruction 'chits'
Emergency procedures eg fire, hospital
Operating instructions for office equipment
Standard letter formats
Instructions to patients eg extractions, anaesthesia, care of
    dentures, orthodontic appliances
Reference lists: dentists' practice and home phone numbers,
    consultant clinics, community dental clinics, doctors

Table II Suggested reception duties

On Arrival
Turn off burglar alarm system
Unlock filing cabinets
Check for messages on answerphone, turn off
Open post, allocate to in-trays
Get out appointment, fees received and fees outstanding
    books
Switch on air conditioning and/or radiators as needed

Daily
Make appointments for patients at the desk and by telephone
Operate and update cancellation list
Record FP17s in ledger and post recorded delivery twice
    weekly. Receive payments of fees, issue receipts and retain
    counterfoils, record in fees received book
Reconcile cash daily
Type day lists, 2-3 days in advance
Get patients' cards from current/DTC files, write FP17s
    needed
File treatment cards at regular intervals
Reconcile and bank cash
Secretarial duties, typing and filing letters
Maintain invoices due file

At the end of day
Put away appointment, cash and fees outstanding books
Update and photocopy updated daylist for each surgery
Lock filing cabinets and put keys away
Put correct 'on-call' tape in answermachine and switch on
Close blinds and switch off air conditioning
Secure premises and set alarm system

Surgery manuals

Tables III and IV suggest subjects worth including in a similar manual for use in the surgery. The dentist may use the same surgery day after day, yet expect DSAs to work in different surgeries without notice, taking for granted how much they have to remember. Written instructions reduce the need to carry mundane details in the mind, releasing it for more stimulating activity. For example, personal variations from manufacturers' recommended mixing procedures can be recorded in the manual, and DSAs can add notes on different materials and techniques.

Reviewing and updating manuals should be entrusted to particular members of staff, who should see that the manuals are re-typed or re-written frequently enough to keep them tidy. Writing manuals does require much thought at the outset, but once set up, they become invaluable.

Table III Suggested topics for a surgery manual

Start-up procedure eg day list, switch-on
Shut-down procedure eg cleaning, switch-off
Clinical procedures eg endodontics, surgical needs
Materials techniques eg mixing, uses
Fee scales: NHS fees and calculations, private fees
Miscellaneous DPB fees eg item 30 items and fees
Stock control, ordering
Sterilisation procedure
Hazards and precautions
Equipment information and care

Table IV Surgery start-up and close down procedures

Starting up
Switch on services: electricity, gas, water, airlines
Fit handpieces, run to blow out oil
Rinse glutaraldehyde off 3-in-1 nozzle and fit to syringe
Rinse glutaraldehyde off burs
Dry and place in bur stand
Check surgery is clean and tidy
Check towels etc
Wipe all working surfaces with glutaraldehyde
Update day list against appointment book for late changes
Chase up any laboratory work needed today and not yet
    received
Get trays of instruments from steriliser
Get RA machine if needed (refer to day list)

Closing down
Check that tomorrow's laboratory work has been received
    (refer to day list)
Ensure all instruments are sterilised ready for tomorrow
Clean out aspirator with aspirator sterilising solution
Clean all burs and all 3-in-1 syringe nozzles in ultrasonic
    cleaner and store in glutaraldehyde
Lubricate handpieces and sterilise in autoclave
Wipe down all working surfaces in glutaraldehyde
Raise operating chair to maximum height for cleaner
Secure windows
Close blinds
Compressor: drain off water from cylinder and trap
Switch off services

Communication within the practice

The introduction of practice and surgery manuals enhances the understanding of routines and procedures, but many messages pass daily between dentist, staff and patients. These too are less ambiguous if in writing.

The apparently simple day-to-day messages from surgery to receptionist (and back) are vital to the success of the practice. The receptionist cannot remember verbal instructions from several surgeries while also dealing with a patient and answering the phone. She must have no doubts about the appointments she is to arrange if the appointment system is to function in an efficient manner.

Memo pads

A simple small notepad in the surgery provides an effective solution. A further improvement may be to use notepads available from dental stationers, but the ideal may be to draw up your own and have pads printed by a high-street copy shop (see over).

All practices evolve their own abbreviations. For example '3 x 30 5a, 2 x 30 fills' may ask for three half-hour appointments with the hygienist for prolonged periodontal treatment and two half-hour appointments to place restorations. Information available from the practice manual may supplement this.

For example, if the laboratory takes 2 weeks to make a chrome cobalt skeleton, the manual should advise the receptionist that appointments for special tray impressions and try-in should be 2 weeks apart. In this way the task of making a complex series of appointments is reliably delegated.

Dentist . . . . Patient . . . . NHS/PT

To Sign: NHS Part 13A . . . . PT consent . . . .
NHS Part 13B . . . . NHS Part 14 . . . .

Return Card & Form . . . .

Fee quoted £ . . . . : . .   To Pay: £ . . . . : . .

Appt: 10 . . . . x . . . .   Remarks:
20 . . . . x . . . .
30 . . . . x . . . .
40 . . . . x . . . .
50 . . . . x . . . .
60 . . . . x . . . .
. . . . x . . . .
. . . . x . . . .

Put on Canc List . . . .

Pro forma: surgery to reception memo.

Only the planned appointment dates should be written on treatment cards by the receptionist. Writing memos on them detracts from treatment notes and fills the cards more quickly, making them thicker so they take up more file space.

Staff meetings

For many dentists, regular staff meetings are an essential aspect of their management style, for others such a notion is totally terrifying. Time is money! What is there to discuss? Fear of staff takeover—the reasons given are many and diverse. While dentists should not use a technique with which they are uncomfortable, they should examine their approach to staff supervision and ask themselves whether it is adequate.

If there is a good working relationship and no need for special staff meetings, all is well, but the dental team will often benefit from open discussion away from the pressures of the practice. An informal setting, such as a 'working lunch' or an evening meeting at home, can be just as useful as a formal meeting in the practice. Whatever the venue, the staff should be able to look forward to playing a rewarding part in shaping their practice.

The meeting is not an opportunity to reprimand. When such an unpleasant task is necessary, it should be carried out as soon after the incident as possible and be restricted only to the person(s) involved. Staff welcome constructive direction, but resent being harangued. Dentists should avoid dominating the meeting so as to encourage everyone to speak out.

There may be some topics which are not to be discussed and it would be well to make these known. These topics would not be the same for every dentist, but 'taboo' subjects might include contracts of employment, the selection of new dentists or new staff, and equipment purchase. On the other hand, things said and attitudes struck at staff meetings should not be disregarded. They can provide important indicators to relationships or practical problems that might otherwise pass unnoticed.

Praise is always welcomed, as there is rarely sufficient time or opportunity in the working day to show adequate appreciation and encouragement. Staff should be invited to submit topics for discussion and the agenda prepared and circulated by the dentist several days before the meeting. This simple courtesy leads to more productive discussion.

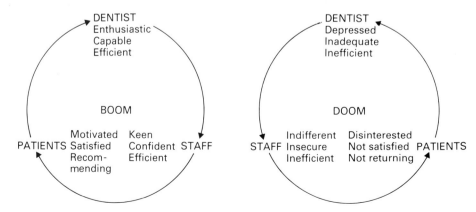

The elation and depression cycles. The role of team leader is demanding; difficult to shoulder if the cycle is anti-clockwise but extraordinarily gratifying when working clockwise.

Just as it takes time to prepare the practice manual, it takes time to develop the practice meeting, and like the manual, it becomes an indispensable tool which benefits everyone.

Staff training

Staff meetings can include wider issues than just day-to-day management. Regular meetings should lead to a reduction in the need to discuss problems and should present opportunities to introduce fresh information.

Involvement can be stimulated by occasionally asking a member of the team to carry out some research and report back. For example, a DSA might be asked to look out information on alternative materials, or a receptionist to think through an improved desk routine or to collect literature to assist in the purchase of a new typewriter.

The increasing availability of video tapes and specialist courses provide opportunities for the team to assess new products, to promote discussion and to learn new techniques. People from outside the practice or from other professions can be brought in to give talks. A colleague might talk about a technique not used in the practice, a teacher discuss infants' attitudes to dentists, or a community health promotion officer be invited to describe his or her work.

Some meetings would need to be held during the normal working day, causing the practice to be temporarily closed. Some courses could necessitate giving staff time off work and reimbursing fees and travelling expenses. Meetings and courses held in the evenings or at weekends would require staff to give up their own time. Despite these costs and sacrifices, the improvement in knowledge, attitude, efficiency, awareness, individual self-esteem and team spirit, far outweigh the disadvantages, bringing benefit to both personnel and the practice.

The main method of staff training continues to be by explanation, demonstration and supervision of day-to-day work within the practice, and the dentist continues to be responsible for seeing that the necessary standard is achieved and maintained. The task is made much easier by manuals which experienced staff can use to teach the trainee, confident that they accurately represent what the trainee should be learning.

Leadership

In order to manage the practice and organise its team, a dentist needs to display leadership. All managerial initiatives —or training, innovation, monitoring performance, maintaining standards and discipline—stem from the dentist. He or she either inspires success and attracts patients, or emanates an air of failure, generates staff dejection and patient rejection. Success breeds success, failure breeds failure.

The role of team leader is a demanding one (see the elation and depression cycle). Dentists need to develop and maintain their leadership skills, just as they do their clinical expertise, by reading and attendance at suitable courses.

Further Reading

Howard, W W. *Dental practice planning.* St Louis: C V Mosby Co, 1975 (hardback).

Harris T A. *I'm OK—you're OK.* London: Pan Books, 1969 (paperback, ISBN 0-330-23543-5, £2·95).

Stubbs D R. *Assertiveness at work.* London: Pan Books, 1986 (paperback, ISBN 0-330-29231-5, £2·95).

Blanchard K, Johnson S. *The one minute manager.* Glasgow: Fontana, Collins Books, 1986 (paperback, ISBN 0-00-636753-4, £2·50).

18

Reception and Office Organisation

M. D. Wilkinson

The reception area is central to efficient administrative procedures and is the embodiment of the style of a practice as regularly seen by every patient. Planning a reception office is not difficult but it should be done carefully to ensure that it is efficient and cost effective.

A split-level reception desk.

Reception staff perform many tasks such as appointment booking, telephoning, typing, filing and writing and so may require several workstations. Comfortable chairs are essential so that energy is not wasted by standing all day. Suitable office chairs cost from £70 to £180†. Because the patients also require a working surface, it is necessary to decide whether they will be required to stand or to sit. If patients are seated, a conventional office desk would be suitable.

If patients stand, the work surface for the staff can be either at standing height with high office chairs, or at normal seated height with normal office chairs, creating work surfaces at two levels: at 100-125 cm (39-48 inches) high for the patients and at about 72 cm (20 inches) for staff. This provides an opportunity to design shelves and pigeonholes under the higher surface, so that reception paraphernalia is out of patients' view but immediately to hand for the receptionists.

The two-level reception desk shown can be easily constructed by a carpenter. If the patients' surface were to be set even higher than shown, at the maximum height of about 125 cm, the pigeonhole system could store large objects such as a typewriter or a computer, as well as smaller items such as stationery and telephones, to form a multi-task workstation for a single receptionist.

The provision of additional work surfaces for other tasks offers postural variety and makes room for more personnel and expansion. Receptionists need about 2 m of work surface to avoid getting in each others' way.

Some time spent browsing through office equipment brochures for inspiration and around DIY stores for ideas on materials will generally lead to a better understanding of the possibilities.

For work surfaces up to 54 cm (21 inches) deep, kitchen work-tops with post-formed front edges are often suitable. Deeper surfaces can be achieved by the use of chipboard or blockboard covered with a durable surface such as Formica or even vinyl flooring, edged with wood, plastic or metal strip.

It is worth establishing the standard sizes in which the materials are supplied. For example, most sheet materials such as Formica are based upon a maximum of 244×122 cm (8×4 ft). Designing a unit which marginally exceeded standard sizes could involve undesirable joints, unnecessary wastage and extra cost.

Kitchen and bedroom wall and drawer units sold in self-assembly kits made from melamine-covered chipboard are well suited to the creation of office (and surgery) units, particularly as they can be cut very easily before assembly.

Melamine-covered chipboard can also be obtained as Contiplas, which is available in several colours from 15-76 cm (6-30 inches) wide and up to 244 cm (8 ft) long. The melamine surface is scratch-resistant but is not so durable as Formica.

Chipboard is also available covered in several natural wood veneers, but whereas the melamine coated type is ready to use, the wood veneers require further finishing and varnishing and are less durable.

A large practice needs a large reception desk with several receptionists allotted different tasks and each receptionist will need her own workstation. Separate tasks might be accounts and payments, appointments for specific dentists to avoid sharing appointment books, and file placement and retrieval. Some tasks may be combined; for instance, a typist may also be required to operate a switchboard and a computer. Staff

---

† All costs include VAT.

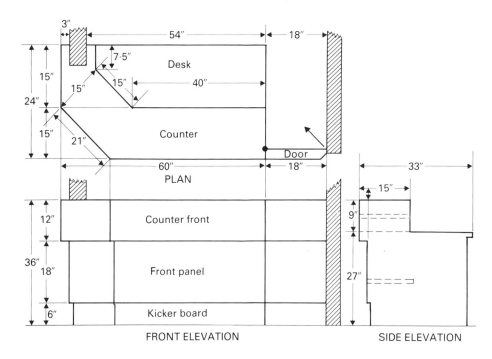

A carpenter's drawing of the desk.

should be trained in every aspect of reception work to provide flexibility and variety.

Filing systems

Treatment record cards take up considerable space—about one metre for 300 cards. The average dentist completing about 2500 courses of treatment per year with the Dental Practice Board has 3000-4000 record cards in his 'active' filing system and so needs 10-13 m of storage.

Shelves are used for a lateral filing system.

The simplest storage system is the 'lateral filing' system consisting of wall shelves on which cards are racked in a row, like books. This arrangement is cheap to make, does not protrude very much and requires no additional space for opened drawers, but it can require a large wall area.

Conventional office suspension-file drawer cabinets offer a reasonably priced, secure system, housing two rows of FP25a cards per drawer. Longitudinal mid-line drawer dividers are usually available from the suppliers, or can be constructed easily from hardboard sheet. A four-drawer unit provides 4·8 m (16 ft) of FP25a file space and costs about £120 if made

A standard filing cabinet system. Note the contrast between the files on the right and the files on the left, where a blue insert tab shows DPB approval is pending, a red insert tab shows that payment is outstanding, and the alphabet is sub-divided by metal window tabs.

from steel, or about £330 if made from wood-veneered chipboard. There are smaller metal filing cabinets available which hold a single row of dental cards but, at about £45 for 40 cm (16 inches) file length, they are less cost effective.

Rotating systems hold more cards than conventional cabinets in the same floor area, but cost even more. For example, a double tier system of 90 cm (36 inches) diameter costs about £480, giving a file length of 10 m (35 ft), and a four-tier file costs about £860 for 20 m. Staff often prefer to

A rotating system holds more than conventional cabinets.

walk between rotary files instead of rotating them, and this, of course, rather defeats their purpose.

Aids to filing

Filing is a constant task, involving many hours each week. Efficiency can be considerably improved by making use of several simple measures:
● The surnames and first names are boldly written at the very top of the cards.
● The alphabet is broken down into groups of 15 to 20 cards by dividers, AA, AD, AT, BA, BE.
● The number of subfiles is kept to a minimum.
● The inactive cards are removed regularly.
● There is adequate filing space to avoid the cards being too tightly packed.

Two or three dentists in one practice could require the receptionists to operate a system containing 9000-12 000 cards, so the risk of misplacing a card is very high. To overcome this, subsidiary filing, colour-coding and numbering systems are usefully employed, often together in the same filing system.

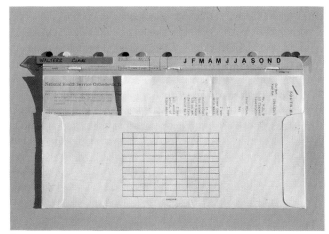

Color Master suspension file wallet.

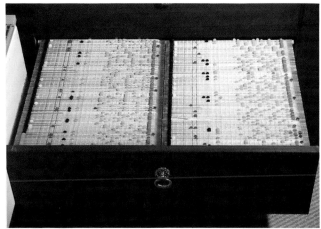

Color Master suspension systems offer special plastic tags which are pushed into slots in the suspension spines.

Subsystems

Many practices operate this method because it seems simple to divide the cards into a separate system for each dentist, and/or a separate system for three or more different categories, such as (a) under current treatment; (b) completed; and (c) inactive ('deads', not returned for, say, 2 years).

Medico-legal adice is that treatment cards should be kept for a minimum of 7 years. A child patient can bring a case against a dentist up to 25 years after the treatment. Inactive record cards are generally removed from the reception system to a 'deads' subsystem located away from the reception area.

Theoretically, the creation of subsystems in the reception files should reduce the likelihood of misfiling and speed up retrieval. However, if a card is misfiled, a search through every subsystem may be necessary before it is found. In a practice of four dentists, each with his own system divided into three further subsystems, there would be a total of 12 subsystems. Although the use of two or three subsystems may be justified there is no merit in setting up a set of subsystems for each dentist.

Colour-coded systems

It is not necessary to physically subdivide a filing system as previously described. Identification of dentists or patient categories can be achieved by adding colour-coding to the cards, filed together in a single system. There are three basic methods of coding:
● Coloured self-adhesive tape or stickers are the simplest means but lack versatility as they cannot be removed easily without damaging the cards.
● Coloured metal tags which clip on the top of the file cards can be added and removed easily, but they can fall off, tend to bend the top of the card and are a little expensive.
● Insert tabs provide a very cheap method of colour coding, are easily added and removed. They are cut from coloured sheets of card to fit inside record cards with the top protruding above the file card by 1 cm (see over). The broad base prevents the insert tab from falling over inside the card. By preparing insert tabs of varying shapes, the tabs can be made to protrude at different positions if required.

In practice, several methods may be employed together. For example, current cards may be subdivided from completed files to speed up the daily routine. A coloured label stuck over the top edge of each card may identify which

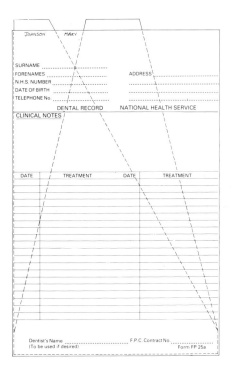

Treatment card with insert tab.

dentist the patient sees, while coloured insert tabs may be used to identify different categories such as 'under current treatment', 'account outstanding', 'DPB approval awaited' and 'exempt from NHS charges'.

The use of the National Health form FP25a treatment card is not mandatory and, provided they carry similar information, other card designs are acceptable. There are some very elegant systems commercially available. Whereas NHS cards are provided free, commercial systems are not, but their greater usefulness makes them worth considering.

Reference numbers

By allocating a number to every patient, record cards can be filed in numerical order regardless of the name. The drawback is that a master index book of names and numbers must be prepared and kept up to date for the times when a patient forgets his number. However, a computer can be programmed to print out the index and to identify patients by both their names and their numbers.

Regardless of the degree of sophistication of any file system, careful labelling and accurate filing are always essential.

Patient recalls

Colour-coded systems are well suited to a recall system using different coloured tags to represent the months when patients are due for a reminder. There are several methods of recalling patients: advance appointments, telephone recall and postal recall.

The first method is the cheapest and is very effective provided that patients remember their appointments, but it reduces flexibility in the appointment book. It is very efficient for rapid screening examination of patients whose dental health is good. Blocks of appointment time can be reserved for such check-ups at intervals as short as 5 minutes each, but higher risk patients and new patients should be allocated longer appointments at a different time.

Telephone recalls are expensive and usually only feasible if there is more than one phone and several receptionists so that one person can concentrate on the task. Details of both home and work phone numbers must be entered on record cards and several attempts may be needed before the patient can be contacted. Some patients may not have a telephone. On the positive side, because the appointment has been agreed with the patient, the failure rate is usually low.

Postal recalls are the least convenient to patients as they have not been asked whether the date and time suggested will suit them, and recall appointments should be planned at times known to be convenient to most patients. Writing out recalls is a tedious task, but it can be eased by asking patients to write their address on the reminder envelopes. These are marked with the recall month and filed in month order in a special recall filing system until they are used. Some practices ask patients to pay the postage.

Postal recalls can be computerised in a permanent database, making it possible for letters and even address labels to be printed out automatically.

When deciding whether or not to operate a recall system the level of patient demand should be assessed. Only if demand is low, can cost and effort usually be justified.

Office equipment

Telephone systems
Telephone systems are becoming versatile and inexpensive. They should be capable of connection to at least two outside lines, connecting every surgery to an outside line, paging between surgeries and between surgeries and reception, connecting an answering machine, and connecting a cordless phone.

By combining all these facilities in one system the need for complicated intercom and light signal systems is eliminated.

Cordless telephone
In a small practice which employs only one receptionist, a cordless phone can improve that person's interest and usefulness considerably, as she can be less deskbound. They cost from £80 to £200.

Telephone answering machine
A suitable machine should have two recording systems: one to record and play back the announcement tape, and the second to record messages received. Several announcement tapes can be compiled and interchanged to deliver different announcements, and the incoming message tape should be at least half an hour long to record incoming calls one after the other.

Other useful facilities are a tone-pager, which allows the owner to telephone the machine and make it play back all the messages it has recorded, message and tape counters, message received indicator, and telephone conversation recording facility. An out-of-hours message telling patients which dentist is 'on call' is an important goodwill function. Answering machines cost from £100 to £200.

Typewriter
Suitable electronic typewriters with a variety of daisywheel prints are readily available for around £300. The good

impression created by a letter produced on a good typewriter is well worth the cost and gives the typist a sense of pride. Letters should not be written by hand if dentistry is to be regarded as a businesslike profession.

A good wordprocessing program, computer and printer may supplant a typewriter, but more expensive typewriters offer such advanced facilities as electronic memories, which store a number of standard letters that can be typed automatically, a spell checker, and the option to connect to a computer to make the typewriter perform as a computer printer.

Electronic cash register

The cash flow of the average dentist is increasing and receptionists are likely to find the careful recording and safe-keeping of monies received increasingly difficult. Cash registers offer improved security and their till-roll print-outs enable transactions to be cross checked against the daily fees book.

Calculator

Calculators are invaluable for working out patients' fees, but they are often rather small. A good desk calculator should be large enough to have well separated keys of about the same size as those on a typewriter. The display should be angled towards the operator and bright enough to be seen easily. Functions should include a % key and a memory facility.

If a cash register is not available, a calculator with a paper till roll will speed the lengthy process of cross checking. Price with till roll is £30-£120. An internal transformer or an external mains adaptor eliminates the need for batteries, however rechargeable batteries are a versatile alternative.

Photocopier

This is a 'bonus' item of equipment which is not essential but brings surprising benefits. It is useful for copying the day list direct from the appointment book or from a typed master list, for copying patient information leaflets, questionnaires, professional articles for reference and so on. A second-hand office copier capable of producing 10-20 copies per minute should cost less than £500. Smaller copiers with lower printing rates, some offering interchangeable cartridges so that different colour inks can be used, cost around £500 new.

Further reading
Fallowfield, M. Keeping tabs on patient files. *The Dentist* 1988; **5:** 44, 47.

Color Master filing systems: Safeguard Business Systems Inc, Centurion House, Gateway, Crewe, Cheshire CX1 1XJ.

19

Appointment Systems

M. D. Wilkinson

At the heart of any dental practice is the appointment book. There cannot be a practice which does not depend upon appointments, yet many dentists seem to be unaware that they can control their appointments system. Practitioners may adopt an existing system which may no longer be appropriate and receptionists can remain oblivious to the pressures and burdens systems impose upon the dentist.

Appointment systems exist to organise patients' needs, to maximise dentists' efficiency and to deliver optimum service. It is for the dentist to decide how to arrange the day's work, paying due regard to staff and patient demands, in order that he can best deliver his skills.

Inflexible time rules
'There are never enough hours in a day' and 'I'll fit that in somehow' are often-heard expressions because people rarely acknowledge that time is limited and rules all life's plans.

Before an efficient appointment system can be established, two essential rules must be recognised:
● First rule: treatment time is fixed
● Second rule: surgery time is fixed
At first sight these rules may seem unreasonable, but they simply state the obvious: that a job of work takes a certain length of time and that a day's work is a fixed unit of time.

Treatment time is fixed
Everybody has a personal average pace at which tasks are best performed and dentists are no exception. Although there are occasions when a task turns out to be more simple or more difficult than anticipated, everyone has his or her own optimum speed of working. Done slowly, the job becomes tedious; done too quickly, energy and skill are dissipated before the day has ended.

It is essential to establish personal working 'norms' before being able to decide how long appointments should be. Over a period of several weeks, the DSA should record how long different treatment procedures take, including the turnround time between patients. At the end of the period, different average times for each procedure will be evident (see below).

An act of discipline is then required. The dentist must accept the findings of this exercise without permitting any intuitive modifications, so that when the patient presents for examination, the DSA can consult the average times and very quickly advise the dentist how much appointment time should be allotted. If the case appears to be difficult, additional time can be added, but the averages should never be reduced matter how easy the case may seem. The appointments required are indicated to the receptionist and she, too, must not adjust them. No one should be allowed to disregard the first rule.

The receptionist's task is to take the different treatment times and fit them together within the working day, which leads on to the concept of . . .

Surgery time is fixed
Everyone works within certain times. If the working day starts at 8 am and finishes at 6 pm for example, all work is to be done within those times. Disregard for them leads to some kind of personal sacrifice as work encroaches upon personal time, and inevitably creates more stress. By defining the working day, it is possible to work proficiently.

Planning an appointment system
Extra demands such as emergencies, phone calls, over-runs and rest breaks occur every day and cannot be regarded as an unimportant afterthought: an appointment system must allow time to deal with them. Before designing a system, the style of the practice and the demands of the patients should be considered, as the appointment system is a powerful tool in shaping treatment strategy and practice goodwill.

The dentist who wishes to cater for patients whose dental awareness is low will need to offer a system which makes attendance easy by offering convenient hours of opening and long periods for emergency treatment. Setting aside other factors, the dentist can use his availability to encourage the groups he is keen to treat. Children should be seen after school. Their mothers would prefer to be seen while their

Treatment	Time			Total
	in	start	finish	
PBGC Prep.	0900	0905	0934	36
6mod7mo8b AM	0936	0938	1001	26
3b2d1mi AE COMP	1002	1003	1026	28
Imp P/P	1030	1033	1044	15
5mod6mo AM	1045	1047	1109	25
Fit 1 PJC	1110	1111	1130	21
Insp 2XR New Pt	1131		1145	15
Consult Tx Plan	1146		1200	15
X732/3 Fit F/F	1201	1205	1226	25

Note that the total time includes the time taken to change patients.

Pro forma: a completed work study.

children are at school, and people in business can be encouraged by offering appointments outside normal office hours.

Emergency periods

Emergency periods avoid rushed treatment and hasty decisions, limit disruption of routine appointments, provide buffers during the day, offer opportunities to do non-clinical tasks and give a better service to the patients.

Providing emergency treatment places enormous strain upon a busy day; leaving a section of the diary free removes much of the strain and offers the patient a better service. Desperate emergencies such as traumatised teeth can still be seen instantly as appointed patients will always accept that their treatment takes second place in such situations.

Emergency periods mid-morning and mid-afternoon can provide extra 'buffers' in addition to the natural lunch time and night time breaks in the system which allow dentists to catch up if appointments fall behind time. If there are no emergency cases, such routines as a tea break, phone calls and case notes can be dealt with.

Instructions to the receptionist should state that emergency periods are not to be booked more than 2 to 4 days ahead, and the periods should be clearly defined by a highlighter pen as shown below. The times of emergency periods should be established with regard to local needs, but if scheduled at the beginning of each session their buffer function is lost. They should always occur at the same time every session so that patients become aware of the facility. There are three systems in general use: the precise-length, the fixed-length and the reservoir systems.

The precise-length system

By obtaining a variety of standard times for different procedures, as already described under the first time rule, it is possible to book a series of prescribed appointments of differing durations. Because the treatment timings have been accurately determined, the dentist is able to work at a steady pace with fewer empty periods with no work to do. The advantages can be summarised. It:
- offers the patient a better standard of work
- greatly reduces stress on surgery staff, particularly the dentist
- provides an even work tempo
- reduces the frequency of empty periods
- avoids congestion at reception and in waiting areas by smoothing the workload of the entire practice
- runs to time, avoiding patients waiting
- helps to make patients punctual because the system is punctual.

Naturally, there are some disadvantages. It:
- takes time and effort to set up
- requires disciplined routines to maintain
- may make it necessary to explain to certain patients why they cannot 'just pop in' to suit themselves.

Fixed-length system

In such a system the appointment lengths are fixed arbitrarily and so are less closely related to the working time. It is simple to operate and no work study preparation is needed, but it has disadvantages.

Some appointments are too short, so treatment over-runs. Some appointments are too long, wasting time. Patients who

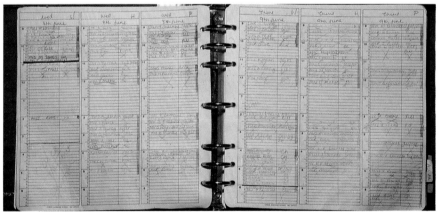

Loose-leaf binding has several advantages. Dentists can tailor a book to suit their needs. One book can be used by several dentists and the length of the working day is more variable. Only one book is needed for many years as old pages are removed and new ones inserted. However it is initially more expensive than conventional books. Binder and pages must be very durable.

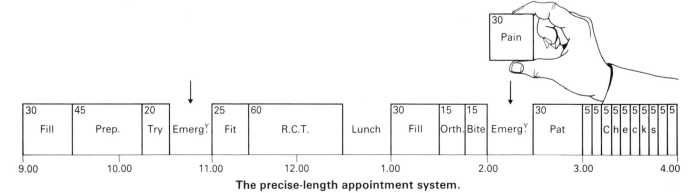

The precise-length appointment system.

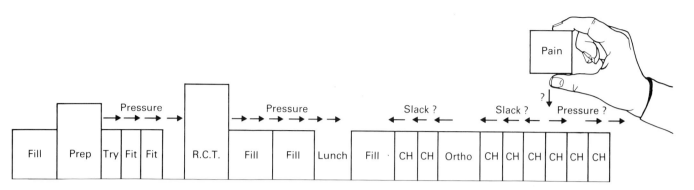

The fixed-length system.

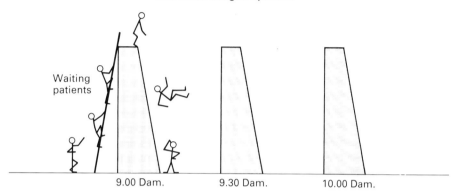

The reservoir system.

are not seen on time have to wait and patients kept waiting learn to arrive late.

With this system, all check-up and treatment appointments might be allotted 15 minutes, on the grounds that some treatment can be done at the check-up if there is no patient already waiting (hardly a scientific basis for careful prescription), or that treatment taking longer than 15 minutes can be carried out by over-running into following appointments.

Patients who are kept waiting become worried and irritable and regret having arrived on time. The appointment system breaks down and instead of controlling its workload, the dental team is obliged to work in a haphazardly reactive way. Whenever anybody keeps another person waiting an insult is implied: the person kept waiting is not important. Patients who are regularly treated in this fashion will eventually seek another dentist.

It is worth remembering that every new adult patient is an ex-patient of another dentist's practice.

The reservoir system

Another system is to give batches of patients an appointment for the same time. By assuming that some will come late, others early, and others not at all, the system evens out the different times needed to treat those who do attend, and all is well in time for the arrival of the next 'batch' of patients. As patients are usually overbooked, it ensures that the dental team is never idle. Although often used in hospital outpatient clinics, this system has very little to commend it, as the waiting frustrates patients and the pressure may lead to rushed treatment.

Appointment books

Some systems are designed around the appointment book, but it should be the other way about. The appointment book should meet the requirements of the dentist. It should also be possible for different dentists working in the same practice to be able to employ different systems tailored to their individual needs.

Once an appointment diary has been introduced into a practice, it is not easy to change over to another style, so you should consider the following features when choosing an appointment book: time units, pre-dated or undated, size, page layout, fixed binding or loose-leaf, special features and durability.

Appointment books simply divide the day into units of time, but they must match a dentist's requirements. For example, using the precise-length system, two appointments of 15 minutes, a check-up of 5 and an appointment of 25 minutes can all be scheduled in an hour.

There should be sufficient space to record the title, initial and surname of each patient, together with a simple treatment code, such as Fill, Sc, Imp, Prep, to enable the team to see the day's workload at a glance.

Although a diary with the dates already printed saves time and error, an undated page allows several dentists to be entered on the same page. In the example shown over, it is possible to see each day's appointments for the entire practice on a single page, and so coordinate appointments with the hygienist and the two dentists. When offering appointments, the receptionist can regulate patient flow through the practice and avoid excessive congestion in the waiting room.

Minimum tools for using appointment books consist of several HB pencils, pencil sharpener and a good quality eraser. The pages should be strong enough to withstand several alterations without wearing thin or tearing. Such close attention to detail is wise. The small cost involved should be compared with the inconvenience, embarrassment or financial loss which could follow damage to a page so serious as to make it illegible.

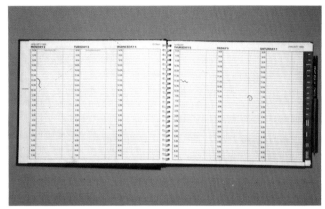 

Bound appointment books are cheap, ready to use, have one week to a page and are not bulky. However, they quickly wear out and may need replacing during the year. They have fixed time units. Each dentist needs at least one book and really two to bridge the year ends.

Dentist:		Date: .... / .... / .... Page:		
Time	Patient	Treatment	Fee NHS	Fee PT

Example of a day/list day book pro forma.

Day lists and day books

From this information a day list is compiled for each surgery, perhaps by photocopying the appointments page, more usually by typewriter carbon-copying. By working only two or three days ahead, the likelihood of last-minute changes is reduced, yet it allows time for treatment cards to be retrieved from the filing system and for NHS FP17 forms to be prepared.

A day book for each surgery is recommended as it provides a permanent record of the patients seen and the treatment provided. An important secondary function is to monitor the output of each dentist and it may even be used as the agreed basis for the remuneration of associates. (The FP17 must be completed as accurately as the day book, otherwise the value of the work claimed for may be less than the value of the treatment carried out.)

The task of writing the names in the book can be simplified by designing a standard day list form which also serves as a day book, even to the extent of including cash columns to record the fees due, as seen above.

Computerised appointment systems

A computer is only as clever as the program in it, and understands only what it is told. There are good computerised systems, but in general computers are unable to read an appointment diary or to coordinate facts as quickly as a competent receptionist. If a manual system has not been used well, a computer will not bring any advantages. It could even lead to a phenomenon known as 'computer-aided disaster'.

Admor appointment books: Admor, Barnham, Sussex PO22 0EW.

Ash appointment book: Claudius Ash Sons & Co Ltd, Casco Wing, Summit House, Cranborne Industrial Estate, Potters Bar, Hertfordshire EN6 3EF.

Cottrell appointment book. Cottrell and Company, 15-17 Charlotte Street, London W1P 2AA.

20

Treatment Plans and Cost Estimates

M. D. Wilkinson

Many dentists find it difficult to discuss charges with their patients. Improving the presentation of treatment plans could make it easier.

Failure to understand an exchange is often the fault of both the initiator and the recipient of the message, but it is invariably the initiator who is held responsible. This apparent injustice is extended further: when a service is purchased, it is the supplier who is blamed, not the consumer; when staff relations are involved, it is the employer who is blamed, not the employee. So it is essential that dentists communicate with staff and patients clearly.

Why tell the patient?

Increased consumer awareness demands that dentists tell their patients the treatment proposed and how much it will cost. This gives patients an opportunity to discuss their treatment and to appreciate the value of professional skills.

The improved simplicity of the NHS charges structure will make NHS estimating easier, but, in the past, the advantages of preparing as full a treatment plan as possible have often been ignored. With very few exceptions, people expect to receive an estimate from a contractor or supplier, who usually waits until asked to proceed with the work or service. In the meantime, potential customers are able to consider whether they want the service, and whether it represents good value, and can seek further estimates. Dentists are being thrust into this market place, and many patients now expect to be able to choose whether they want to be treated or not.

Written estimates

To achieve complete understanding, estimates should be in writing. For straightforward courses of treatment, a summary of the treatment proposed showing the anticipated cost is adequate information. This can be worked out quickly by the dentist, DSA or receptionist and given to patients at the examination, before they agree to make any further appointments. A copy of a suitable pro forma simple estimate (see right) is filed in the treatment card. The routine use of fee charts and ready-reckoners considerably speeds this operation.

Treatment plan and estimate

When more involved treatment is envisaged, a combined treatment plan and cost estimate should be provided, whether under NHS or private contract. The time spent in planning and costing can be recovered from the private patient by charging a fee, but the only return under the NHS contract is the probability that the patient will accept the plan and the near certainty that payment will follow.

Treatment planning is the proper procedure following diagnosis. The dentist should use all the information available: accurate charting, radiographs, study models, previous treatment *and* what the patients have to say—listening and understanding their wishes, as well as

considering their needs. The time and effort required is repaid by the reduced likelihood of an unexpected problem during treatment, and by the increased involvement of the patient.

The planning and costing of complicated treatment should be done away from the pressures of daily practice, and the final plan and cost estimates typed and sent to patients so they can understand them, appraise them and make a decision, at home, away from any perceived pressure in the practice.

The extent of a plan and estimate can be adjusted to suit the situation. The basic method of formal presentation is the letter, describing in simplified terms what treatment options are suggested, why and how much they will cost. The medical right/left convention is usually best reversed or explained, but over-simplification should be avoided, so that the patient can understand the technicalities involved and appreciate the value of the treatment.

A further aid to presentation is to prepare and duplicate a

TREATMENT ESTIMATE

for..

accepted for treatment under
NHS/Private contract by

....   Date: ..../ ..../ ....

The cost and treatment advised is shown below. Please note that while every effort has been made to ensure accurate planning, changes may be necessary and so the cost may change.
Payment is required as detailed.
Please bring this estimate to each visit.

Treatment advised	Fee	Paid
Total	£	£

Payment due at first visit: £.............
Payment due at each visit: £.............

Pro forma for simple estimates.

series of standard illustrations and text of various routine procedures—crowns, bridges, dentures and endodontics, for example—which can be added to appropriate letters. The easiest way to produce illustrations is to photocopy drawings or photographs from textbooks and journals (with due regard to copyright). The line drawing below was prepared by tracing a photograph from a textbook. Several tracings were modified with fine felt tip pen, pasted onto sheet paper together with a typewritten text, and photocopied.

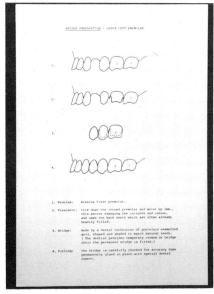

A line drawing to illustrate a treatment plan.

The ultimate in presentations is to take Polaroid photographs of the patient's own teeth. This never fails to impress and always adds meaning to the estimate.

A finishing touch is to prepare a cover page and bind all the pages together in a clear plastic cover with a slide-on plastic spine. The care employed in its preparation will not go unnoticed. Patients will use it to discuss the plan with relatives and friends, or possibly other dentists. Even if they decide not to proceed with the treatment, they usually will not throw away the plan, but refile it for later reference.

Finally, there should be an invitation to make an appointment to discuss the plan and a copy filed with the treatment card.

It is unrealistic to present only a plan without including a cost estimate, and for this, the fees scale is essential.

NHS fee scales

The NHS scales are laid down in Determination I of the annual 'Statement of Dental Remuneration'.[1] Many practices use this as their only source of information, but quick reference charts, such as those published by Admor or Cottrell, considerably speed up the completion of FP17s and the fee calculations.

Another useful aid is the hand-held electronic calculator and programs (next page) marketed by Filax, along with its optional printer that prints out a summarised estimate. Whenever the NHS fee scales are revised, new reckoners are published.

Private fee scales

As the level of NHS charges rises, patients are more likely to wish to discuss alternative private options. A list of private fees should be prepared for their reference. As dentists are free to determine their own level of private fees, there are no standard private fee scales available.

Every dentist offering private treatment should prepare a fee scale each year. Without one, the stress of rapid mental calculations is added to clinical demands, with the further risk of underestimating or over-pricing. The former produces low profits, the latter loses patients.

When preparing a private fees scale information should be gathered on:

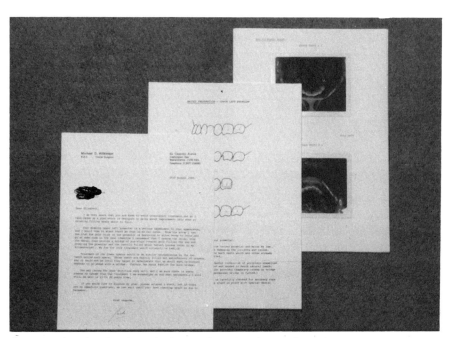

A very extensive treatment plan and cost estimate for private treatment.

The Admor NHS fees reckoner.

The Filax fees reckoner.

- personal average treatment times
- material costs
- laboratory costs
- hidden costs, such as consultations and treatment plans
- desired technical standard
- the fees charged by other dentists in the same area.

Fee costing

This aspect of treatment planning worries many dentists. There is neither shame nor surprise in this. Whereas a businessman enjoys making a profit, dentists often get 'price fright' and quote too low a fee in order to avoid the patient's refusal.

There is no harm in the patient refusing to take up the treatment offered. In the same way that shop sales assistants do not expect to sell to every person who enquires the price of their wares, dentists also have a passive role: that is, to instil into each patient a desire for the treatment they have recommended, so that when the patient later decides that the time is right, he or she will return to conclude the 'sale'.

When costing treatment, it is essential to recognise two important principles: the dentist is worthy of his hire, and a refusal to buy is not the end of the patient's interest.

What is my time worth?

Many emotive things have been said about the value of dentists in the UK in contrast to around the world, but there is only one real value. That is the price that people are prepared to pay for dental treatment. Before an objective scale of private fees can be drawn up, the dentist needs to establish how much his services are worth per hour, his/her 'time value'.

There are a variety of methods available to work this out, but one of the simplest approaches is to examine historical earning capacity and, from that, to produce a projected target earning rate. This is done by extracting the following figures from the annual accounts and from the diary for the corresponding period:

- gross fees received
- net income before tax ('taxable income')
- total hours spent practising chairside dentistry.

These figures are then applied to the general formula:

$$\text{Income/chairside hour} = \frac{\text{income before tax}}{\text{chairside hours worked}}$$

Example I

John Cusp's annual accounts for 1987/88 show:
Gross fees received = £58 000
Taxable income = £23 000
Over the same period his diary shows that he worked an average of 7·5 hours per day for 5 days per week. He took the statutory 9 days for national holidays, 20 working days for annual holiday and 5 working days for sickness and courses a total of 34 working days' absence. He therefore worked for 226 days, or 1695 hours. His historical earning rates for that year were therefore:

$$\text{Net income/hour} = \frac{23\ 000}{1695} = £13 \cdot 57$$

$$\text{Gross fees/hour} = \frac{58\ 000}{1695} = £34 \cdot 22$$

This formula can also be used to determine forecast targets for the year ahead.

Example II

If John Cusp wishes to earn £25 000 taxable income in the year 1988/89 (an increase of 9%), assuming that his hours will remain unchanged, he can uprate the previous year's figure to calculate how much he will have to 'net' per hour:

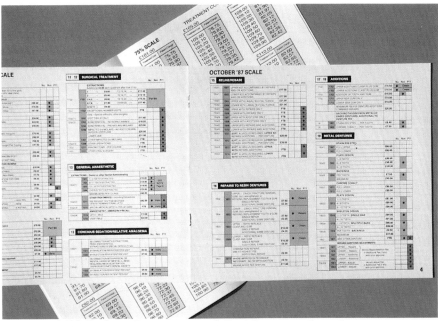

Cottrell's NHS Easicast fee computer & Easicharge ready reckoner.

Net income target = $13 \cdot 57 \times 1 \cdot 09 = £14 \cdot 79$/hour.

If his overheads ratio remains the same

Gross income target = $34 \cdot 22 \times 1 \cdot 09 = £37 \cdot 30$/hour.

These figures represent the average time values overall for both John Cusp's NHS and private treatment.

Private fees: how much should I charge?

In order to see how much NHS and private treatment were worth per hour, the accounts are again examined to establish how much of the gross fees was derived from each source and, from the day book, how many hours were spent providing NHS and private treatment. From these facts, the hourly historical earnings and the target earnings for NHS and private treatment can be calculated.

Example III

Time spent providing private treatment	= 250 hours
Private fees earned	= £12 000
Gross private fees earned/hour	= £48

If gross income is to increase by 25%:

Gross target fees/hour = $48 \times 1 \cdot 25$	= £60

In order to formulate a complete scale of private fees, this hourly gross target is applied to the dentist's individual work rate (recommended in chapter 19).

Example IV

Average time to place posterior composite:

$4 \rfloor$ mo	= 40 mins
Average gross fee at £60/hour	= £40

When laboratory fees are involved it may be necessary to decide whether or not to calculate target rates net of laboratory fees. In practice this can be ignored if the proportion of laboratory overheads is less than 15% of the annual gross overheads. The argument is that this 'overcharge' represents a modest additional return for the greater expertise applied to items involving laboratory techniques.

Example V

Provision of a private porcelain crown:

Appointment time taken to prepare	= 40 mins
Appointment time taken to fit	= 25 mins
Therefore chairside time value at £60/hour	= £65
Laboratory fee	= £35
Private target fee	= £100

It is up to individual dentists to decide whether the fees they ultimately charge will be strictly in accordance with their scale of fees, or whether they will charge according to actual time and laboratory costs.

Can I earn a living under the NHS?

The NHS scale of fees is determined by the Department of Health each year, based upon the Government's decision on the recommendations of the annual Doctors' and Dentists' Review Body. This is an exercise in historical accounting and forward projection aimed at providing the Target Average Gross Income (TAGI) and Target Average Net Income (TANI) for the coming year for the average dentist, working at the average pace for the average number of hours per year, with average overheads, providing the average mix of treatments. These averages are translated into the NHS scale of fees by the Dental Rates Study Group (DRSG).

TANI and TAGI are thus directly linked and, although dentists can alter their clinical and business methods and their overheads in an attempt to match the 'average', by undertaking to treat a patient under the NHS, they accept the general constraints of the in built average profit margin (Table 1).

To decide whether he or she can provide treatment under the NHS each year and meet targets, the dentist must establish his or her individual work rates and then relate them to the year's NHS fees.

Example VI

Time to place a molar mod	= 20 mins
1988/9 NHS fee item 0604	= £12 · 40
Gross fee earned per hour	= £37 · 20

Table 1 DRSG settlements — the 4-year record

	1985/86	1986/87	1987/88	1988/89	1989/90
Target average net income[1]	£20 083	£21 345	£23 220	£24 920	£26 915
Increase over previous year		6·3%	8·8%	7·3%	8·0%
Increase over previous 4 years				24·1%	25·0%
Average gross income[2]	£46 579	£51 747	£56 258	£64 823	£65 622
Increase over previous year		+11·1%	+8·7%	+10·2%	+5·8%
Increase over previous 4 years					39·2%
Increase in all items RPI					
Over previous year		+3·2%	+4·0%	+6·0%	+7·8%
Over previous 4 years				20·5%	22·6%

1 TANI figures for 1985/86 and 1986/87 are adjusted to take account of phasing in implementation. All TANI figures include income from pool-funded schemes — seniority payments in all years and vocational training payments in 1987/88 and 1988/89.
2 Actual gross payments are shown for 1985/86 to 1988/89. For 1989/90 a target figure is shown. All gross figures include seniority payments and vocational training receipts, as for TANI above.
(Source: British Dental Association)

Example VII

Provision of bonded porcelain gold crown under the NHS:

Appointment time for preparation	= 30 mins
Appointment time for fitting	= 20 mins

Therefore, applying the gross target fee of £37·30/hour from example II:

Chairside gross fee	= £37·30
Laboratory fee	= £34·00
Total	= £71·30

Compared with:

1988/9 NHS fee item 1032	= £71

These figures closely match John Cusp's gross hourly target fee, calculated in example I. If all the other items of treatment needed by this patient matched his NHS 'rate-value' criteria, John Cusp might well be happy to proceed. But if any overhead were to change during the year, such as an increase in laboratory costs, he would be wise to re-examine his profitability before offering further NHS contracts.

It is essential to recognise that it is only at the initial examination that the dentist recommends private or NHS treatment, and that the patient is required to accept before treatment can commence. Only those dentists who have calculated their own financial targets and have prepared their own private fee scales will be able to make treatment plan and cost proposals which are accurate and financially viable.

Reference

[1]NHS General Dental Services 'Statement of Dental Remuneration, Determination I'. Issued annually through FPCs by the Department of Health.

Admor fees reckoner: Admor, Barnham, Sussex P022 OEW.
NHS Easicast fee computer and Easicharge ready reckoner: Cottrell and Company, 15-17 Charlotte Street, London W1P 2AA.
Filax electronic fees calculator: Filax Ltd, 65 Tredegar Square, London E3 5AE.
Polaroid (UK) Ltd: Ashley Road, St Albans, Herts AL1 5PR.

21

Financial Management

M. D. Wilkinson

Dentists enjoy solving problems, applying their special skills and serving their community, but how often is the financial reward less than deserved? Could this have been avoided? Was the fee costing at fault, was the treatment plan unrealistic, or was there a simple omission in practice administration? Here are some basic administrative procedures which can help to promote efficiency.

Primary efficiency

This means getting it correct at the outset, so we can stop worrying about it. When tempted to make an inadequate decision or to apply less than our best to a clinical situation, we should remind ourselves that, some day, someone will have to put it right. Will that be us or another dentist? Will the patient have suffered in some way? In the short term only we know whether we have provided clinical work which is as perfect as we can get it, but failure to insist upon primary efficiency within our practice soon becomes obvious to everyone. Inefficiency has to be paid for in many ways: personal frustration, financial cost or loss, in having to work harder to counter the extra burden and sometimes even having to repeat procedures.

Some administration of practice finances is invariably delegated to staff and procedures must be correctly followed if the delegation is not to be regretted.

Fee calculation and recording

The first point at which money can be lost is when calculating the fees. The chairside DSA is invariably responsible for making the entries on the treatment cards. An omission or error here leads to the loss of a fee which will probably go unremarked for ever. Good DSA training and subsequent supervision are essential.

Some dentists then delegate the transfer of the items of treatment from the treatment card to the FP17 or the private account to a secretary or receptionist, but time and geographical separation from the chairside may render it impossible to refer back to resolve any queries. Such confusion can be avoided by asking the DSA to complete the NHS dental estimate form FP17 as she enters the record card at each appointment. This also gives her the opportunity to discuss form details with the dentist.

By the end of treatment, both the record card and the estimate form are fully and accurately completed, ready for final calculation by the receptionist, who is able to collect the fee without any delay. If the simple estimate pro forma (shown in chapter 20) is in use, the stage payments are just as easily calculated and charged at each visit.

Book keeping

All businesses need a set of books to record receipts and expenditure and, while dentists may regard the need to complete form FP17 as an imposition, they are spared the problems of collecting VAT endured by most businesses.

Fees received and due from all sources earned by all the dentists in the practice, and where and how that cash has been lodged, should be accurately recorded. This is a simple requirement which should be kept simple.

The receptionist is usually responsible for fee collection and will need:
- fees received book
- fees outstanding book (can be combined)
- bank paying-in slips
- receipt book
- DPB book listing all FP17s sent to the Board

The fees received book is best ruled into columns to display both an analysis and a detailed record of the transaction (see over).

As the level of patients' NHS contribution rises and the amount of cash handled increases, the risk of theft also rises. Payment by cheque and credit card markedly reduces this risk. Staff should be fully instructed in the procedures to be followed when accepting cheques supported by cheque cards and payment by credit cards. Credit cards cost the practice 3–4% of the payments received.

Cashing up and banking

The receptionist should total up the monies received at specific intervals, such as every evening or at the end of each session, and seal them in an envelope marked with the date and sum enclosed, before placing it in the safe. Later, when cashing up for banking, each envelope is opened in turn and the contents checked against the book entries before opening the next one. This procedure isolates any errors that may have occurred on any one day.

How frequently practices bank the receipts will depend upon the rate of cash receipt, the security of its safe-keeping and the requirements of the practice insurance policy. There are usually limits to how much money can be kept on the premises, depending on the type of safe used and how much money can be carried in transit. Fees should be banked by the end of each week to avoid keeping cash on the premises over the weekend. Banking daily ensures that the bank balance is kept as high as possible, reducing overdraft charges, and minimising loss due to theft, but it may be inconvenient.

Make your bank work for you

Taking cash from the practice receipts is accompanied by several risks. Errors and omissions may pass unnoticed. Theft by personnel is easier and it reduces your ability to refute accusations of fraud by the Inland Revenue. These risks can be substantially reduced by insisting that all monies received are paid into the practice bank account.

By paying all income into the bank, the bank unwittingly acts as accounts clerk: it records each payment or counter credit in the practice statement, and should be instructed to

DATE	Rec No.	PATIENT	DENTIST 1 NHS	DENTIST 1 PT	DENTIST 2 NHS	DENTIST 2 PT	DAY TOT	TO BANK
1·5	205	G Johnson	26·35				26·35	
	206	J Smith			150·00		150·00	
	207	M Erikkson	4·05				4·05	
		E James		545·00			545·00	
	208	A Baird			16·85		16·85	
		W Littlej'n				35·00	35·00	
	209	R Williams	73·58				73·58	
	210	H Lloyd	33·00				33·00	
		DAY TOTAL	136·98	545·00	166·85	35·00	883·83	
2·5		etc						
		WEEK TOTAL					– – –	
		MONTH TOTAL DENTIST 1 DENTIST 2 NHS PRIVATE					– – –	

Typical fees received book (note that this is self-checking: the horizontal totals equal the sum of the vertical totals each day).

supply this monthly. The paying-in book carbon copy retained by the practice lists every cheque banked and so provides a breakdown of counter credits, and tallies with entries in the fees received book.

If petty cash is required, a practice cheque should be written and substituted for the cash taken from the fees received, or the cheque taken to the bank and cashed there. This is better than taking cash and making cross-entries in the fees received book. The practice accountant and tax inspector will be happier with this method of cheque substitution than with cash removal and staff will know that all transactions are 'above board', and opportunity for theft is small.

Using this system, all cash withdrawals and all payments are made by cheque, so that they appear in the practice bank statement. In practices where it is the daily custom for the dentists to take the fees they have each earned, very careful records should be kept by the practice and by each dentist, so that the Inland Revenue can be in no doubt. Banking these sums so that the dentists' personal bank accounts match the practice entries will similarly offer some protection from Inland Revenue suspicion.

Once the Inland Revenue decides to investigate, it immediately searches back 6 years and is empowered to look even further back. Such investigations can take many years, costing a good deal in additional accountant's fees, which are borne by the practice regardless of the eventual outcome, and are extremely worrying.

The NHS Terms of Service require that patients are given form FP64 receipts or similar, and for the same reason there can be no disputing the sense of this.

Details of all cash expenditure should be recorded in a simple petty cash book and supported in detail by shop receipts.

Recording and collecting fees outstanding

Debt accumulation is best avoided by training all personnel (dentists included) to infer that fees are due at specific points in the treatment. This may be payment in full or in part at the beginning, at each visit or at completion. This inference, reinforced by appropriate notices and fee estimates, is usually all that is required to secure payment.

The procedure should be strictly followed to ensure that patients recognise that they are dealing with an efficient business, determined to secure its dues.

The few patients who have not cleared their debts by the end of treatment are entered into a fees outstanding book. This need be no more than a book with an alphabetic index, drawn up in columns to show date of debt, name, fee due, date of accounts sent and payment received. Card index systems similarly prepared are also suitable, but they are more susceptible to failure if any cards are removed.

Account reminders must be sent at regular intervals, such as fortnightly or monthly. Increasingly blunt warning stickers can be added, or the wording of the accounts can become more aggressive, and the third account should warn that steps will be taken to press the claim. If payment has still not been received, the account should be referred to a debt-collecting agency, which will deduct a fee of 25% to 33% from debts successfully recovered. For debts in excess of £50, it may be worth taking the debtor to the small claims court, which, although charging an initial fee, will award that fee and other costs, such as interest, to the successful claimant.

Dentists should not be afraid to put their dentist-patient relationship to the test by telephoning the patient, as there may be a legitimate reason for non-payment. Personal visits are probably unwise.

Debts can best be monitored by the preparation of a monthly summary. In addition to the fees outstanding book, treatment cards should also be marked or tab-indexed as a back-up system and to prevent a debtor receiving further treatment and credit without authority.

DPB book

In its simplest form, this book comprises a list of all FP17s sent to the DPB, including those sent for approval. Its main function is to list them so that should any be lost in transit, they can be rewritten by reference to the FP25a treatment cards. Envelopes full of forms can go astray and it is worth using registered post or recorded delivery.

If forms are lost, the DPB not only fails to pay, but also requires the forms to be recreated in full, including patients' signatures.

A more sophisticated use of the DPB book is to record the value of treatment, so that it can be checked against the monthly schedules to identify any errors.

Forms should be sent to the DPB at frequent intervals, in quantities which are not too heavy for the envelopes. Batches of about 70 represent a reasonable workload for practice staff to list and ensure a steady flow of payments from the Board.

Income-expenditure analysis

The best insight into the performance of the practice can be obtained from a monthly analysis prepared by the dentists themselves. If this task is delegated to staff or an accountant, we may not be as closely aware of the costs, incomes and trends. The items detailed in the analysis columns should be those we wish to see in order to understand the business. An accountant will be able to prepare the year-end accounts from such an analysis so easily that our efforts are repaid by a reduction in accountancy fee.

The expenditure side is usually more complicated and involves more entries per month than the income side, and it is better to design this before buying a ledger, which should be of sufficient size to take one month's expenditure on a single page. There are many options available from business or dental stationers. If payments are made by practice cheque or credit card in the third week, most of them will have been presented by the end of the month, so statements can easily be checked against the analysis.

Expenditure is twice net income

In current NHS dental practice, if the level of income tax averages 25% and overheads are 60% of gross, expenditure is twice net income. This is shown by the example:

Gross fees	100	
Less overheads	(60)	
		40
Less tax at 25%	(10)	
Net fees		30

equal to 50% of overheads

The control of overheads is thus twice as effective as working harder. Unfortunately, there is a limit to the extent to which overheads can be reduced without jeopardising the business.

Expenditure can be divided into fixed and variable overheads. Fixed overheads are those commitments which generally continue throughout the life of the business and, although loans and mortgages may be paid off eventually, they do not vary from day to day. They are generally paid by regular instalments. Running costs also tend to be a continuous charge on the working practice, although some reduce or cease when the practice is not working and there are opportunities to negotiate them from time to time.

Consumable costs are directly linked to work: no work, no consumption. Their uptake and cost can be controlled through wise treatment prescription and material selection, while good management can regulate the value of stock held.

Stock control

Unit cost, bulk discounts, consumption rate, shelf-life, reorder level, maximum and minimum stock holdings, unit bulk, storage space, delivery time—these are the factors involved in keeping a stock of supplies.

The average practitioner spends £3000 to £5000 each year on dental materials and often leaves their reordering to his or her DSA. Without guidance, the DSA will tend to order exactly the same items as before and will tend to over-stock to avoid accusations of incompetence. Optimum stock control is achieved only by use of a stock control system.

The minimum information needed to order an item is item

Income and expenditure analysis book (Admor).

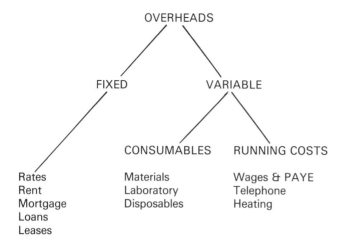

Diagram of overheads.

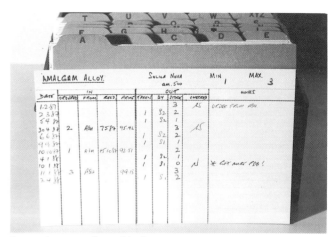

A stock control card index system.

name, unit as packaged, and the supplier. However, better control is achieved if unit cost, bulk discount rate, delivery time, shelf-life and consumption rate are also known.

Most efficient control is achieved by using this information to determine minimum order level, maximum reorder level and the best-priced supplier.

There is no sense in ordering five cases of 1000 cartridges of lignocaine 2% if the consumption rate is 1200 per year and its shelf-life is only 2 years. Any discount for the bulk order will be outweighed by cartridges wasted as they lose their potency. While 1000 LA needles at £5 per 100 is not too great a financial strain, if the consumption rate of diamond burs is only four per month, holding 100 burs at £1·50 each is unnecessarily extravagant. A commitment of £150 represents 37 months' supply.

To determine the consumption rate, each surgery needs a surgery order book in which to record every item taken into use, together with date, item and quantity.

The minimum stock level is determined by considering the delivery time, the consumption rate and the frequency of stock reordering.
Consider for example, periapical radiograph films:

Delivery time	= 5 days
Packaging unit	= 150 films
Consumption rate	= 150/week
Minimum stock level	= 2 packets

The maximum stock level must reconcile bulk discounts with cost outlay, bank interest charges, storage space, shelf-life and consumption rate.

There are two methods generally used to administer stock control: shelf stock control and a card index system.

A shelf control system is a physical count of the stock to be seen, often requiring considerable shelving, so that the entire stock can be seen together, as in a supermarket. A special stockroom is the ideal, but if located in a general-purpose room, the shelves should be enclosed. The shelves should be shallow so that stock can be counted easily. A stock card is fixed to the shelf in front of each item, detailing how many units should be there; for example: Baseline, packs 2. Stock levels can be seen at a glance, brand duplication avoided and regular stock checks can identify items no longer in use.

A card index system may be filed in a filebox or in a loose-leaf book. To facilitate reordering, the stock card shows the product name, source, cost, and minimum and maximum stock levels. A record of details and dates of every stock movement allows the rate of consumption of each item to be determined. The stock levels need to be physically checked only once or twice per year because the cards can be used instead. If, for some reason, it is not possible to provide a single stockroom or area, a card index system makes it possible for stock to be stored in several places within the practice and in a variety of ways—on shelves, in cupboards, drawers or bin systems.

When restocking, new packs should always be placed behind the older ones to ensure correct stock rotation.

Further reading
Greenway J. *Finance for the dental practitioner*. London: British Dental Journal, 1986 (paperback, ISBN O 904588 11 4, £2·50).

Analysis Books: Admor, Barnham, Sussex P022 OEW.

22

A Computer in the Practice

M. D. Wilkinson

It is possible to buy a computer system, a printer and business application programs, all well designed for use in any office, for about £700. This brings computerisation within the reach of the smallest practice. This chapter outlines how business systems can help in practice and provides a guide to the purchase of systems which can be upgraded to use specialised dental programs.

Where to start

The beginner would be well advised to start by introducing a computer for basic use at the business office level. Even if eventual total dependence upon a computer is contemplated, expertise and knowledge gained on the way will be beneficial and, if the initial system ultimately proves to be unsuitable for larger applications, it could still continue to be used in a less demanding role elsewhere, such as for stock control.

What can computers do?

Modern computers are immensely powerful tools, capable of performing far more demanding operations than those usually employed in business. Computers can be applied to dental practice at three levels: business applications, specialised dental programs, and business and dental programs together. Business application programs such as wordprocessing, database and spreadsheet programs are bought ready to use, and are generally intended to be 'customised' by the user. For example, wordprocessing packages contain no pre-existing letters, and databases require the user to draw up his own forms for the files.

Specialised dental programs are intended to provide several applications in a single package, in many instances derived from business programs that have been tailored to the requirements of dental practice.

The cost of the equipment and the programs should be offset by the advantages. If computerisation enables existing staff to undertake more work when an extra receptionist would otherwise have been recruited, then the outlay can easily be justified.

Business applications

It cannot be assumed that a computer will create order from chaos; it is the user who is in charge.

Some business applications which may improve dental office procedures are:

Wordprocessing for letters and text

This is a function of which all computers with the right software are capable. Typists can see the text on the screen, arranging and correcting it before printing it. Text can be stored in the memory and printed out at any time and the stored text can be modified. Text such as letters can be combined with address files to produce personalised letters and labels with different addresses. This is known as 'mail-merging'. Different type-fonts can be used in the same text and spelling can be automatically checked and corrected.

This series was written on Wordstar Express, a typical, low-priced wordprocessing package costing £80. Without it, I would have had to type out, cross out and rewrite the text many times, a task considerably eased with a wordprocessor.

Databases and spreadsheets for accounts

Database programs create files and lists of information such

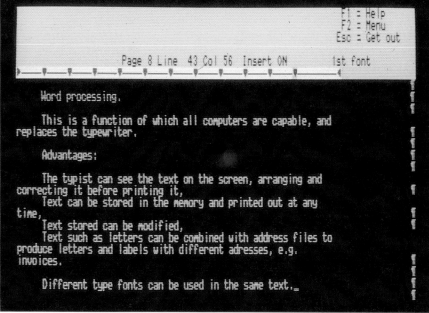

Word processing in Wordstar Express, a typical, low-cost package.

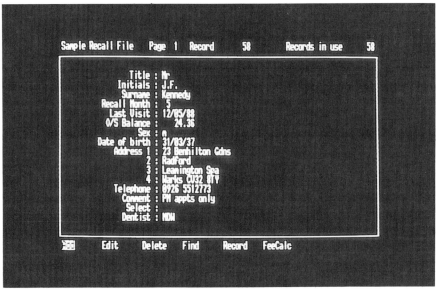

A patient's file from Recall-PC, a dental application program.

as patients' details, debts, fees collected and stock control. More powerful programs are capable of cross-referencing information between several databases, and all can print out the data in 'reports' or summaries. An example of a database file from a specialised dental program is shown above.

Spreadsheet programs enable pages of tabulated numbers or words to be prepared, such as accounts and financial analysis. Anything which is recorded in a book can be recorded in a computer, but, unlike a book, a spreadsheet can be set up to carry out calculations automatically within the page. For example, a column of figures can be totalled, and the total carried forward to another set of figures with which it can be added, multiplied or divided.

An extremely helpful spreadsheet function is 'what if?'. Given a spreadsheet totalling the practice analysis, the question 'what if I can reduce my overheads by 4%?' can be instantly answered by multiplying all overheads by 96%, whereupon all related calculations, such as profit, will be seen to change to fit the new situation.

An important drawback to single-application programs is. that the user is obliged to stop working with one program before he can move to another, which he then has to start up. This is not only inconvenient, but, more significantly, information cannot readily be transferred from one application to another. Integrated programs overcome this.

Integrated programs
In such programs, several applications are brought together, so that the user can move from one to another with relative ease, while drawing upon the same information.

Wordstar Express integrates its wordprocessing and database applications to provide the facility known as 'mail-merging': information taken from the database can be inserted into a standard letter created in the wordprocessor to produce 'personalised' letters to all the people listed in the database. Thus we can create a file of patients, detailing fees outstanding and recall dates, which can be mail-merged to print out letters individually addressed to the patients, demanding settlement of outstanding accounts or inviting them to attend for recall examinations.

Ability Plus is an example of an integrated suite of many programs: wordprocessor, database, spreadsheet, graphics, presentation and communication. Information and text can be moved about between the applications. For example, graphics can produce a pie chart from figures contained in the spreadsheet, pie charts and spreadsheet figures can be included in text in the word-processor, and the whole report can be mail-merged with a database file and printed out as a personalised report.

'Presentation' organises a sequence of information and pictures, possibly suited to patient education, and in conjunction with special encoding equipment known as a 'modem'; 'Communication' permits data to be exchanged via the telephone system with another computer. Ability Plus costs £230.

For beginners, wordprocessing is probably the most useful application. After mastering wordprocessing, users will move onto the address list database and from there into the mail-merge facility (a very useful function for society secretaries). Anyone interested in making financial analysis easier will want to purchase a spreadsheet, and eventually a dentist will probably want to introduce some degree of computing into his or her practice.

Dental applications
Dental programs are designed to be used in dental practice without users needing specialised computer knowledge. They should offer full integration of many of the above applications, be easy to use and have rapid information retrieval and calculation times. Ranging in price from about £600 to £3500 for single user application, their cost directly reflects the number and power of the applications included and their degree of 'user-friendliness'.

It may be difficult for prospective buyers to decide what applications they will need, and they may have to pay for applications which they may never use, but dentists should avoid failing to include applications which they may use later. An 'ideal' dental package may have the following features.

Wordprocessing and mail-merging
From within the dental program it should be possible to gain access to conventional wordprocessing. If this feature is not

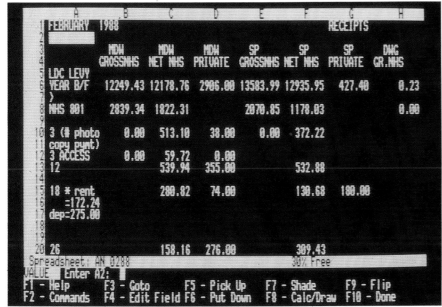
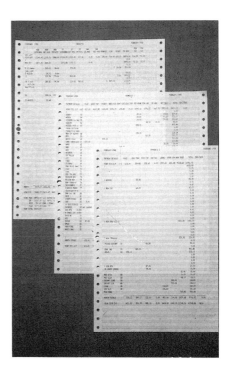

The screen view of a spreadsheet program (Ability Plus) applied to the income/expenditure analysis already seen prepared by hand in chapter 21 of this series. The same analysis is shown printed out *(right)*

powerful enough, the possibility of replacing it with a preferred program should be investigated before purchase, as the wordprocessing portion is often not customised and therefore may be interchangeable.

Patient records and financial control

This application should be able to record all patients' personal details in a database file, together with other information such as the treatment last provided, what it cost, the amount still owing, the last attendance date, the next appointment date, when the next check-up is due, and special information such as exemption, age group and dentures.

It should be possible to combine this data with the integral wordprocessor and to print out accounts, reminders, recalls and cancellations. It should also be possible to produce reports and analysis of incomes received, such as DPB schedules, private fees, the earnings of each dentist and of the practice.

Fee calculations

Some dental programs offer automatic fee calculation. The user types in the treatment and the program shows and records the fees due. These programs need updating when fees are revised.

Treatment charting

There are dental programs which can show a charting grid on the screen for completion by typing the notation into the keyboard. To make use of this, it is necessary to have a terminal in each surgery, in addition to a central computer in the reception area. The immense amount of disk storage memory needed to store the charting slows down the speed of access to the information.

The cost of several computer terminals and the large disk storage memory can be considerable. Earlier preoccupation with programs offering complete charting of the patient's dentition and recording treatment has therefore subsided.

Appointments

Theoretically, computers should be well suited to making appointments, especially if they have been programmed to offer such options as finding the first Friday when the patient could see both the hygienist and the dentist after 3.30 pm. But despite their incredible speed, computers are still slower than a well-trained receptionist with a well-designed appointment book, who sees the whole day or the week at a glance and has a 'feel' for the bookings already made. The risk of losing such invaluable information is also a worrying problem (see back-up, below).

Direct DPB linking

The DPB is currently conducting trials of systems able to accept direct input of information normally recorded on the FP17 to its computer from a computer terminal in the practice. Electronic banking methods could eventually see the DPB paying directly into the practice account on the same day that the computerised claims are sent by the practice.

Back-up

The more information held in a computer system, the more important it becomes to ensure that the information cannot be lost. Data can be lost by total or partial power failure, by equipment failure and by accidental erasure. Loss can be minimised by back-up copying of data at frequent intervals, by controlling power surges and by providing a reserve power supply.

Only frequent copying of the records can protect data from accidental erasure, so that the copies can be used to put the information back into the data storage memory. The cheapest method is to use one of the commercially available back-up programs to copy the data onto floppy disks. Back-up copies should be taken once or twice a day if the system is in constant use, to restrict loss to data recorded since the last back-up. The copies should be kept in the security of a safe, away from strong magnetic fields.

An alternative is to use a 'tape streamer' system, in which the data is backed-up onto a special cassette tape. Tape streamers cost about £400.

The minimum requirement to regulate power surges is a special plug or socket fitting which contains a device to absorb sudden changes in the electrical supply. This costs about £12. In any event, it is unwise to connect any other equipment to the same mains-fuse circuit as a computer. Reserve power systems cost at least £500 and cannot completely counter the risk of a 'crash'.

Maintenance contracts

Good contracts for the maintenance of the equipment and for consultancy support and updating of the programs are essential if the practice is to become fully dependent upon the data stored within a computer system. This can cost from £60 to £600, and represents another cost factor to be considered when moving up from an office support system to a full practice system.

Who to ask

There are four alternative sources of advice and training. You can buy a computer and software to teach yourself or you can seek the advice of a local retailer, a business computer specialist, or a specialist in dental computer systems.

For those willing to learn their way into the subject, current low prices provide the ideal excuse to buy a system. Bought as a practice expense, the first computer should be installed at home, where the new user can learn at his or her own pace from hands-on experience, and from reading the many books and magazines widely available.

Computer retailers may claim that they know all the answers, but their aim is to sell, not advise, and there are very few who understand the needs of dentistry. A local retailer can provide hardware, business software, servicing and local support supplies, but usually cannot assist with the detailed modification of programs to suit special needs. Business computer specialists are more knowledgeable than retailers, but they tend to demand high fees and are not as conversant with dental needs as dental computer specialists.

Dental computer specialists certainly do know what is required. There are several gifted advisers around, many are dentists themselves. They also expect a good fee for their good advice. However, they tend to offer only their own programs and rarely have knowledge of any others. Their experience is often limited to no more than a few dozen installations over the past 5 years and they often insist on packaging the software with a particular brand of hardware, which may be expensive. However, if a practice is to become dependent upon a total dental practice system, with good servicing and staff training, the dental computer specialist is a sensible route to take.

As computer equipment has become reliable and programs easier to use, dental software packages are now becoming available 'off the shelf'. They may not require major disruptive changes or expensive training, and because dentists can select their own hardware and proceed at their own pace, they are likely to become increasingly cost effective and popular.

Choosing a system

The basic hardware items required to run business software are a computer, comprising a visual display unit (VDU), a central processing unit (CPU) and a keyboard, and a printer to print out 'hard copy'.

In order to operate more advanced applications which require large data storage memory, for example specialised dental programs, more equipment may be added, such as a hard disk or hard card, to increase the data storage memory, or a modem to communicate with other computers.

Software requirements are programs, with clear user manuals and floppy disks, and also paper (continuous feed or single-sheet feed), labels (continuous feed) or, as an alternative, window envelopes.

Programs operate in special codes called 'environment or operating systems'. There are several environments; however, most are not compatible with each other. In business software, the two main environments used are MS-DOS (Microsoft Disc Operating System) and CP/M (Digital Research Control Programme for Microcomputers). As they are not interchangeable and need different hardware, it is important to decide which system to adopt.

IBM, the world's leading computer manufacturer, uses a special version of MS-DOS. Computers made by other manufacturers to run IBM programs are known as 'IBM compatibles' or 'IBM clones' and they are generally much cheaper than IBMs. If it is possible that a computer system may eventually run a program which needs a large data storage memory, such as a database of all patients, an IBM or IBM-compatible computer would be better than a CP/M computer because its data storage memory can be expanded easily by the addition of a hard disk or hard card.

CP/M systems are as competent as IBM-compatible svstems for routine business purposes involving less extensive data storage and tend to be less expensive.

The printer does not have to be the same make as the computer. Printers are made in two standard carriage widths, 15 inches and 10 inches, and are based upon three main types of print mechanism, the dot-matrix, the daisywheel and the laser. Dot-matrix printers drive nine or 24 small rods against the print ribbon to form the print. Daisywheels strike each plastic letter of a rotating disk against the print ribbon, and laser printers employ a laser-beam photocopy technique.

Laser printers are extremely rapid and produce excellent print, but cost several thousand pounds. Dot-matrix printers are very much faster than daisywheel printers, but daisywheels produce better print—as good as a typewriter. However, dot-matrix printers can produce a wide variety of type faces in the same text without any need to adjust the printer and many print almost as neatly as daisywheel printers.

Cost

Since the introduction of the Amstrad range of IBM-compatible computers in 1986, the cost of hardware and software has fallen dramatically. Until then, industry-standard business software programs were not available for computers costing less than £3000 to £4000*.

All costs include VAT, correct at January 1990.

An Amstrad PC 1512 SD-CM in a dental surgery, with an Amstrad DMP3000 dot-matrix printer above it.

Prices of the IBM-compatible Amstrad 1512 and 1640 range from £445 for a monochrome single-drive model with 512 kilobytes of random access memory (RAM), to £1100 for a full-colour model with 640 kilobytes of RAM and 32 megabytes of in-built data storage memory. This provides adequate capacity for 10 000 to 15 000 patients' records. A printer costs from £200.

If a system without IBM compatibility is thought to be adequate, the CP/M system is cheaper. For example, the Amstrad PCW 9512 computer system is ready to use for £505, complete with a daisywheel printer, Locoscript wordprocessing, Locospell spell checking and Locomail mail-merge programs.

Computer strategy

As computer technology continues to develop, earlier systems will be superseded by newer, faster ones. The superseded systems will be retained, but their applications may be downgraded to perform less demanding tasks, such as stock control, located away from the reception area.

The costs of computer hardware and software continue to fall, and larger practices may find advantage in purchasing

A Short Glossary of Terms

CPU   Central processing unit. Contains the electronic gadgetry (chips) and the disk drive(s).

Data storage memory   Data input to the computer is recorded either on a floppy disk, or on a hard disk, the size of which limits the amount of data which can be stored.

Disk drive   Usually built into the front of the CPU, but sometimes separate; there may be one or two fitted. Drives spin the floppy disks and read the data on them.

Floppy disk   A flat disk of magnetic-coated plastic (as on cassette tape) enclosed in a plastic or cardboard case. It provides a removable data store of the information needed by the computer. Memory capacity 360K or 720K. Made in non-interchangeable sizes: $5\frac{1}{4}$ inches diameter (in cardboard case) and $3\frac{1}{2}$ inches diameter (in a plastic case) are the most common sizes.

Fonts   The term describing the type. Daisywheel fonts can be changed only by changing the wheel. Dot-matrix fonts can be changed by program instructions.

Graphics   The term describing any non-text information such as graphs or cartoons.

Hardware   All computer equipment.

Hard cards   Similar to hard disks, but of different construction, fitted inside the CPU.

Hard disk   A sealed pack of special disks spun continuously by an electric motor, located either permanently within the CPU, or as an external unit which is plugged into it. Memory capacity is usually 10, 20, 32, 40 or 50MB. Up to 32MB most widely used, costing about £300. Much faster than a floppy disk drive.

Keyboard   A 'QWERTY' typewriter-like unit used to type information into the CPU.

Kilobyte (K)   Approximately 1000 bytes of memory (actually 1024 bytes), where one byte approximates to one letter or space of text. See also megabyte.

Megabyte (MB)   Approximately one million bytes of memory (actually 1 048 576 bytes).

Printer   Connected to the CPU, prints the information onto paper, which is called a 'print out' or 'hard copy'.

RAM   Random Access Memory. The part of the computer 'brain' in which the program is temporarily stored when it is in use. Typical capacity is 512K or 640K. The RAM requirements of a program must not exceed the RAM of the computer, or the program will not run correctly. RAM capacity will increase as new technology is developed.

Software   The program which generates the text seen on the VDU.

Software support   The contractual help provided by the company which supplied the program. May include updates.

VDU   Visual display unit, or monitor. The 'TV' part with the screen, usually stands on the CPU. Can be monochrome (green, amber, grey or white on black) or colour.

more than one computer. The next logical step would be to 'network' them together, so that the same programs can be accessed from different terminals.

Computers: the answer?

Dentistry has become inured to the high capital cost of its specialised equipment. Now that powerful computers are cheap, they can be very cost-effective and of benefit in dental practice. However, computers are only another tool which may enhance good manual systems. They cannot redress inefficiency, and will not replace the need for sound managerial ability.

Further reading

Hill S G. *Computers in dental practice*. Manchester: NCC Publications, 1988 (paperback, ISBN 0 85012 734 3, £14·50)

Joseph St J, Burgess B & R. *The Amstrad PC 1512: A user's guide*. London: Collins, 1987 (paperback, ISBN 0 00 383405 0, £12·95)

Amstrad computer systems and software: Amstrad Consumer Electronics plc, Brentwood House, 169 Kings Road, Brentwood, Essex CM14 4EF.

Ability Plus integrated application software: Migent (UK) Ltd, Harbour Yard Building, Suite 310, 3rd floor, Chelsea Harbour, London SW10 OXF.

Recall-PC dental recall software: Specialist Dental Services Ltd, Westside House, 123 Bath Row, Edgbaston, Birmingham B15 ILS.

Wordstar Express wordprocessing software: Amstrad Consumer Electronics plc, Brentwood House, 169 Kings Road, Brentwood, Essex CM14 4EF.

23

Case Assessment for Orthodontic Treatment

W. P. Rock

Successful correction of overbite and overjet after orthodontic treatment may be judged in terms of incisor overbite and inter-incisal angle. Removable appliances can produce worthwhile overbite reductions, but the final inter-incisal angle depends upon skeletal pattern. It is vital, therefore, to make an assessment of this factor when planning treatment for a Class II malocclusion.

This short series of chapters will consider the selection of those Class II malocclusions which are suitable for treatment with removable appliances. It is not possible to achieve a textbook approach with limited space and the main emphasis will be upon correction of overbite and overjet. The following assumptions will apply throughout:
(a) Well cared for dentition with no teeth of doubtful prognosis.
(b) All permanent teeth erupted, or well placed on radiographs. Panoramic radiography is particularly useful for orthodontic screening purposes.
(c) No severe tooth malpositions. Arches either well aligned or with mild to moderate crowding, such as might be resolved by distalisation of buccal segments, or by removal of first premolars.
(d) No access to cephalometric radiograph equipment.

Treatment aims
The work of Edward Angle has been brought up to date by Lawrence Andrews,[1] who identified orthodontic treatment goals as the 'Six keys of normal occlusion'. The six keys are:
(1) correct relationship of first permanent molars;
(2) correct mesio-distal crown angulation of all teeth;
(3) correct bucco-lingual crown angulation of all teeth;
(4) no rotations;
(5) no spaces;
(6) flat curve of Spee.
It is possible for a skilled orthodontist using fixed appliances to produce a 'six keys' result from all but the most difficult starting point and there is no doubt that such a result is highly desirable in terms of stability and appearance. Removable appliances, on the other hand, can only move teeth by tilting them, so that more modest objectives must be set. Incisor positions must be correct for appearance and function, particularly for the maintenance of overbite stability. This objective should be of paramount importance, yet it is often overlooked or incompletely understood. The principles will be explained in a later chapter. Also, there should be good buccal occlusion with even intercuspation.

Choice of treatment method for overjet reduction
Overbite and overjet may be corrected by several types of tooth movement:
(1) Repositioning of both upper and lower incisors using fixed appliance systems. Fixed appliances have the great advantage of being able to control axial inclination during tooth movement. This makes it possible to reposition the incisors in a way that is not possible with removable appliances.

(2) Reduction of overbite and retroclination of upper incisors using removable appliances. The lower incisor angulation and antero-posterior position cannot be changed.
(3) Alteration of arch relationship using functional appliances.
(4) Alteration of jaw size, shape and relationship by means of surgery.
One of the four methods described above may be used alone or it may be necessary to employ combined treatment to produce the best result. When removable appliances are used alone, careful case selection is essential if acceptable results are to be achieved. Malocclusions may be divided into two groups:
(1) those where removable appliance therapy will produce worthwhile improvement in a reasonable time;
(2) malocclusions where removable appliance therapy should not be attempted since the end result would be unsightly and possibly damaging to the teeth and periodontium.
Although the wishes of the patient and parents must be considered, the desire of all parties to achieve a reasonable result quickly must not be allowed to dominate a treatment plan. As stated earlier, modern orthodontic mechanisms in skilled hands can deliver results of great occlusal perfection and if a less than ideal result is to be accepted, there must be a full appreciation from the outset that this is so. There is an identifiable degree of Class II malocclusion beyond which it is unreasonable to expect removable therapy to achieve an acceptable result; yet every consultant orthodontist, from time to time, is asked to provide a second opinion for a case where such treatment has been attempted and failed to produce expected improvement, or worse still led to disaster (fig. 1). Such failures can be avoided by correct treatment planning and appliance selection.

The production of normal overbite and overjet
The following section will consider those factors that interact to produce normal overbite and overjet and the implications of these effects upon planning orthodontic treatment.

Factors that determine overbite and overjet
All tooth positions are determined by the interplay of three factors:
(1) skeletal factors;
(2) soft tissue factors;
(3) dental crowding.

Skeletal factors
An assessment of jaw relationship is an important step in successful orthodontic treatment planning. Removable

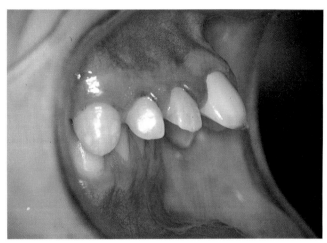

I have a belief this patient is not wearing his removable appliance, judging from progress and breakages.

I wonder if you would consider him for a fixed appliance. There has been some improvement.

Fig. 1 A Class II division 1 maloccusion after unsuccessful treatment with removable appliances, plus the letter of referral.

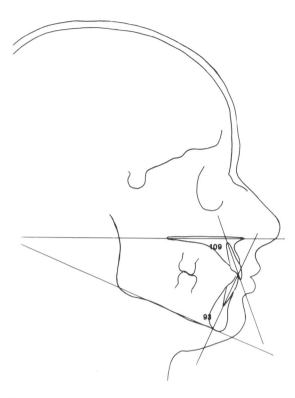

Fig. 2 The tracing of a lateral skull radiograph showing normal incisor angulations.

appliances can only achieve overjet reduction by tilting back the upper incisors and they must not be used in cases where there is more than a mild degree of antero-posterior skeletal discrepancy. Otherwise an unacceptable amount of incisor retroclination will be the result.

Assessment of jaw relationship
Accurate assessment of jaw relationship requires the use of a cephalometric radiograph (fig. 2)[2]. When a cephalogram is unavailable, it is still possible to assess jaw relationship by clinical measurement. In order to understand the method it is necessary first to consider the cephalometric principles upon which it is based.

In most Class II division 1 malocclusions the overjet is due partly to the skeletal base relationship, and partly to the tilting effects of soft tissue action upon the teeth. When a cephalometric film is available, a simulated treatment result is arrived at by means of a prognosis tracing (fig. 3). This tracing is produced by superimposing a second sheet of tracing paper upon the first. The lower incisor is moved down to the occlusal plane without altering its axial inclination. The upper incisor is then retroclined about a fulcrum one-third of the way up the root until the overjet is fully reduced. This method gives a visual impression of the treatment result. If a cephalometric radiograph is not available, the jaw relationship is assessed by two clinical techniques.

Palpation of points A and B (fig. 4)
This procedure is facilitated if the examiner has two fingers of equal length. Those unfamiliar with the method may be surprised at how much more readily a retruded mandible is detected by palpation than by vision. According to the horizontal distance of B behind A, the skeletal discrepancy is described as mild, moderate or severe.

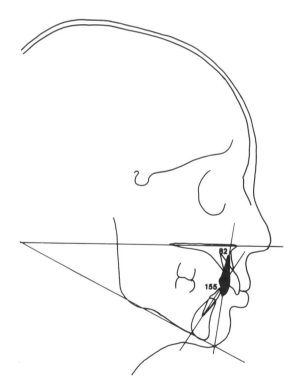

Fig. 3 A composite tracing showing the likely result of overjet reduction by removable appliance treatment. The solid outline is post treatment.

Clinical measurement of incisor angulation
With care it is possible to make a clinical assessment of incisor angulation without the use of radiographs. For this examination the patient is seated in profile to the examiner with the Frankfort plane parallel to the floor. This plane is an imaginary line between the lower border of the eye and the upper edge of the external auditory meatus. Although the

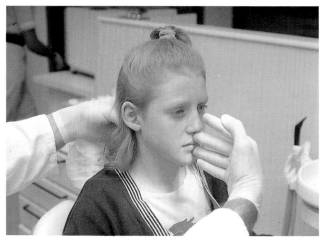

Fig. 4 Palpation of A and B points as an assessment of jaw relationship.

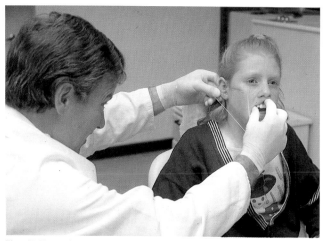

Fig. 5 Use of a perspex template as an aid to measuring incisor angulation.

Frankfort plane is not usually parallel to the maxillary plane, when the cranial base is used as the reference point, it is sufficiently close to be used in place of the maxillary plane and still provide recordings of reasonable accuracy.

The angle between the upper incisors and the Frankfort plane is then assessed. It is helpful to use a perspex template cut to the normal upper incisor angle of 109° as an aid. By sighting past the template it is easy to estimate the degree of proclination of the teeth and from this to calculate approximate angulation (fig. 5). Overjet is measured in millimetres, using a steel ruler, from the tip of the most labially placed upper incisor to the labial surface of the corresponding lower, parallel to the occlusal plane.

Assuming 2·5° of tilt for every millimetre of overjet reduction, it is now possible to calculate the upper incisor angle that will be produced if overjet is to be corrected by tilting the upper incisors. For example:

Original angulation is 119°, overjet = 8 mm.

Overjet after treatment will be 2 mm.

Therefore, the reduction needed is 6 mm or 15° of retroclination, and final incisor angulation will be 119°–15°, 104°.

As a general rule, the lowest acceptable upper incisor angulation at the end of treatment is 95°, that is 14° retroclined from the normal. Further retroclination produces two undesirable effects:

(1) Appearance is poor due to dishing-in of the profile.
(2) Marked retroclination of the upper incisors will produce an increased inter-incisal angle that might allow over-eruption of the lower incisors and produce a traumatic overbite. The normal inter-incisal angle is 131°, the highest that should be created is 145°; above this value the tendency for lower incisors to over-erupt may be inadequately resisted.

Soft tissue factors

Thirty years ago, British orthodontists were very concerned over the action of the lips and tongue in producing malocclusion and in causing relapse of treated orthodontic results. In the 1950s, especially, Ballard, Tulley and others published widely on such topics as 'The tongue, that unruly member'.[3]

The British fixation with soft tissue effects arose largely because, at the time, orthodontics in the United Kingdom was almost exclusively in the realm of removable appliances. In particular, it was felt that, in most cases, it was not desirable to alter the position of the lower incisors by tilting, since relapse would inevitably follow, due to the action of the lips and tongue. The work of Mills[4] confirmed the view that the lower labial segment was positioned in the neutral zone between lips and tongue and that attempts to move lower incisors by proclination or retroclination, over more than a small distance, would be unsuccessful in the long term.

American orthodontists had traditionally taken a more mechanistic view of treatment planning. According to this philosophy, the aim is to reposition the lower labial segment in order to create a better profile. The maxillary teeth can then be positioned over the lowers at the correct angulation to the maxillary plane. This gives very pleasing aesthetic results and avoids production of the dished 'orthodontic face'. Stability is assured by a combination of two factors. First, accurate repositioning of incisors in terms of antero-posterior position and angulation. This requires the use of fixed appliances that can exert control over incisor angulations. Secondly, prolonged retention in order to allow time for the lips and tongue to accommodate to the new tooth positions. Bonded retainers are becoming increasingly popular, especially in the lower incisor region, since they are inconspicuous and well tolerated (fig. 6).

Over the last two decades there has been a unification of orthodontic practice throughout much of the world. In the United Kingdom, fixed appliance therapy is the norm in specialist centres. In consequence, there has been a reduction in the concern felt by British orthodontists over soft tissue influence. It is now realised that with correct planning and accurate treatment the lower incisors will, within limits, remain stable in a new position and that stability of overjet reduction will be maintained by adaptation of the lips and tongue. However, with removable appliances the old rules still apply and the lower incisors must not be proclined or retroclined.

Dental crowding

Surveys suggest that it is more common in the UK population for teeth to be crowded, than to be well aligned. There are several methods by which space may be obtained for

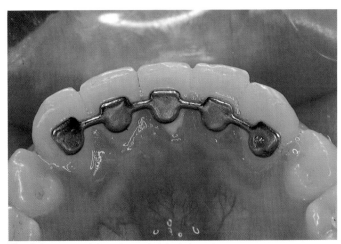

Fig. 6 A commercial ready-made bonded retainer.

realignment of crowded teeth:
(1) extraction of teeth;
(2) arch expansion: (a) lateral;
                    (b) incisor proclination;
                    (c) distalisation of molars;
(3) uprighting teeth that are tilted in the line of the arch;
(4) approximal enamel reduction (interdental stripping).

The amount of space created should match closely the amount of space required to relieve crowding. Excessive space creation will leave residual spacing, a situation no longer acceptable to parents or to the discerning orthodontist. The days are gone when a patient with residual space following premolar extraction should be dismissed with the message that 'It will close in a few years' time'. Such optimism is often unfounded. It is of course possible to close residual spacing completely by using fixed appliances and results are good. However, it is bad practice to have to fit a fixed appliance merely to close space at the end of a removable appliance treatment.

The best guide as to the amount of space required for tooth alignment and overjet reduction is the relationship of the canine teeth. This will be discussed fully in chapter 24. Briefly, if the canine relationship is more than half a unit Class II, the best results with removable appliances are normally achieved following the extraction of first premolars. This facilitates treatment and does not affect the quality of the contacts between the remaining teeth.[5] If only half a unit of space is needed for overjet reduction, as indicated by a canine relationship that is cusp to cusp, buccal segment distalisation becomes the treatment of choice.

Summary

The success of overjet and overbite correction may be judged in terms of the final inter-incisal angle and lower incisor biting position.

Inter-incisal angle

Removable appliances can only tilt teeth and cases must be selected in which the final upper incisor angle will be above 95°. This will ensure that appearance is acceptable and also create an inter-incisal angle that is not increased to a point at which over-eruption of the lower incisors becomes likely.

Biting position

The position of greatest stability for the lower incisor bite is the cingulum plateau on the palatal surfaces of the upper incisors. If the lower incisors are left biting in this region, the chance is minimised that over-eruption of these teeth will produce eventual periodontal trauma following treatment. Proper bite opening is an important part of orthodontic treatment and will be considered at length in the next paper. Unfortunately, correct overbite reduction requires a degree of care in appliance management that is often neglected.

References

1 Andrews L F. The six keys to normal occlusion. *Am J Orthod* 1972; **62:** 296–309.
2 Mills J R E. *Principles and practice of orthodontics.* p 75. Edinburgh: Churchill Livingstone 1982.
3 Tulley W J. The tongue: that unruly member. *Dent Pract* 1965; **15:** 27–38.
4 Mills J R E. The stability of the lower labial segment. A cephalometric survey. *Dent Pract* 1968; **18:** 293–306.
5 Foster T D. *A textbook of orthodontics.* 2nd ed. pp 209–231. Oxford: Blackwell Scientific Publications, 1982.

24

Treatment of Class II Division 1 Malocclusions

W. P. Rock

It is possible, with careful case selection, to achieve good orthodontic results using removable appliances, provided that they are well designed and properly adjusted. The best way to measure progress is to divide the treatment plan into stages, with each stage having recognisable objectives that must be achieved before moving on to the next stage.

It is important to select cases in which removable appliance treatment will produce a result that is both aesthetically pleasing and functionally stable. The role of acceptable incisor angulation in deciding appearance and overbite stability has been considered.[1]

Although orthodontic treatment is a gradual process, it is possible to divide a treatment plan into stages, with each stage having recognisable objectives. The main stages in the correction of overbite and overjet are:

(1) Creation of space to permit tooth movement.
(2) Retraction of upper canine teeth into a Class I relationship with the lowers. Overbite reduction is also begun during this stage.
(3) Overjet reduction. Before this can be achieved, bite opening must be completed so that there is space into which the upper incisors may be retracted.
(4) Retention to ensure stability of the result.

Stage 1. Space creation

Canine relationship is the most valuable yardstick for assessing the antero-posterior space requirement. In order to provide sufficient space for full overjet reduction and correct alignment of the incisors, a Class I canine relationship must be achieved. This is where the tip of the upper canine occludes in the embrasure between the lower canine and first premolar, with the angulations of both canines correct.[2] If the upper canines are normally angulated and the lowers are markedly mesially inclined, the latter must be repositioned mentally before a true appreciation of canine relationship can be gained.

On the basis of canine relationship, the most appropriate method for the correction of antero-posterior discrepancy may be chosen. If more than half a unit of space is needed for overjet reduction and there is no lower arch crowding, the upper first premolars alone may be extracted. During treatment the upper first permanent molars will move into a full unit Class II relationship with the lower teeth and ensure a good buccal occlusion. However, if there is lower incisor crowding in association with the overjet increase, the best plan will be to remove the four first premolars. The removal of first premolars will provide readily accessible space for canine retraction and also help to preserve even buccal occlusion and a correct first molar relationship.

If the canine relationship is less than half a unit Class II and the lower arch is not crowded, buccal segment distalisation is the first choice treatment. This approach will require the use of extra-oral traction.

Stage 2. Canine retraction and the beginning of bite opening

The two objectives are considered together since there is

increased overbite in the great majority of Class II malocclusions. Naturally, if the overbite is not increased, no adjustment is required.

It is wise to fit an appliance before teeth are removed. The design chosen should be simple, comfortable to wear, and strong. The patient must understand the appliance and the way in which it fits. It is counter-productive to fit a complicated appliance, for instance one that incorporates passive springs that are to be used during a later stage of treatment. Such an appliance is less likely to be worn properly than a more simple design and is more likely to produce unwanted tooth movements, especially an overjet increase due to anchorage slip.

If the canines are in the line of the arch, the type in figure 1 is a good standard design. The short labial bow, which contacts low on the labial faces of the upper incisors, provides anterior retention and also helps to guard against anchorage slip by preventing incisor proclination during canine retraction. The flat anterior bite plane will be trimmed as described later.

Before placing the appliance into the mouth, the following points should be checked:
(a) Base plate smooth with metal tags correctly trimmed.
(b) Wirework of the correct gauge, clasps and labial bow 0·7 mm, springs 0·5 mm.
(c) Spring coils correctly placed. A common fault is to find coils placed too far back so that during retraction there is a tendency for the canines to be rotated and pushed out of the arch (fig. 2).

When the canines are outstanding at the start of treatment, further buccal displacement and rotation may be minimised by using 0·5 mm buccal retraction springs sleeved to the coil with soft tubing (fig. 3). The tubing supports the distal leg of

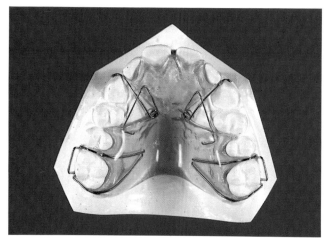

Fig. 1 An appliance to begin canine retraction nd overbite reduction.

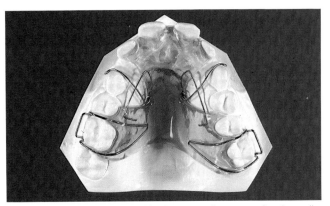

Fig. 2 Canine rotation caused by incorrect placement of spring coils.

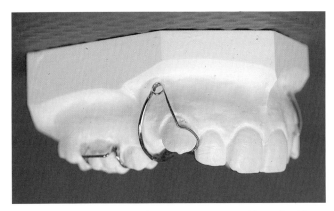

Fig. 3 A 0·5 mm buccal canine retractor, supported by soft tubing.

the spring, whilst the mesial leg has good flexibility. Buccally approaching springs of this type have less tendency than palatal retractors to displace the appliance so that anterior retention is not usually necessary when they are used.

Appliance adjustment

A removable orthodontic appliance requires frequent adjustment and trimming if it is to work at maximum efficiency. Monthly recalls provide the best way to monitor progress. At the time of fitting, the following adjustments are required:

Clasp activation

If the technician has trimmed the model before making the clasps, as recommended by Adams,[2] no activation will be needed. However, undercuts may be engaged more positively by bending the clasp slightly over the approximal contacts at the time of fitting.

Spring activation

Correctly made springs are adjusted by placing the round beak of spring forming pliers into the coil and twisting so that the coil is opened slightly to give around 2 mm activation. The arm of the spring must not be bent, since this alters the point of application and hence the direction of tooth movement. Activation may be checked by placing the appliance lightly in the mouth with the Adams clasps on the molars, but not seating it fully. The passive springs should lie just in front of the cusps on the canines.

Bite plane adjustment

Orthodontic appliances are not normally made on articulated models so that the height and depth of the bite plane is a guess. Unless the guess was inspired, the acrylic will need to be trimmed at the time of fitting. The polished surface of the acrylic is first roughened and an occlusal registration obtained using articulating paper. Height is then reduced by grinding away the marks until the molar teeth are about 2 mm apart with the appliance in place. Finally, the plane is trimmed flat and excess acrylic is removed from the vault of the palate to within 2 mm behind the marks.

Bulky, over-extended bite planes are a common cause of poor appliance tolerance and careful trimming is important. After 2–3 months the molar teeth will re-erupt into occlusion and the bite plane must be built up again

to continue bite opening. This build-up may be achieved by the addition of cold cure acrylic and then trimming as before.

Assessment of progress

The 'ideal' rate of orthodontic tooth movement is around one millimetre per month, so that it will take 6–7 months to complete canine retraction if the teeth must be moved through one unit of space. It is wise to measure overjet carefully at the outset, and at intervals thereafter, so that any tendency for anchorage slip may be detected early. These measurements should be taken in a standard way, for example parallel to the occlusal plane, half way along the incisal edge of the left central incisors. Overjet increase may indicate over-activation of the springs or the need for anchorage reinforcement by the addition of extra-oral traction.

Once the canines are almost fully retracted, the labial bow may be cut off since it has served its dual purpose of safeguarding anchorage and providing additional anterior retention. At the same time, the collets around the incisors are trimmed to a smooth curve, so that the incisors have the chance to spread out. This movement relieves incisor crowding and facilitates overjet reduction during the next stage of treatment.

Assessment of completed canine retraction

Correct canine retraction and bite opening is the base from which overjet reduction is carried out. The natural wish to begin overjet reduction must be resisted until stage one objectives have been fully achieved. If there is sufficient space, it is wise to over-retract the upper canines slightly, since they will soon relapse to a correct position.

Stage 3. Overjet reduction and continuance of bite opening

The most efficient spring for retroclining upper incisors is the Roberts' retractor (fig. 4). Unfortunately, this spring is often made poorly and in a way that reduces its usefulness. A Roberts' retractor should be made in 0·5 mm wire, sleeved to the coils in soft tubing to give support. Coils must be carefully positioned to avoid soft tissue damage (fig. 5). The anterior section should not extend more than half-way across the lateral incisors, otherwise it will catch on the canines in the later stages of overjet reduction. The coil should be midway between the legs on each side, so that the 'front' leg is angled forwards. When activating the retractor, care must be taken not to stress the wire where it emerges from the tubes.[3] Activation of 2 mm should be achieved at the coil by using the thumb to bend the wire around the beak of spring-forming pliers (fig. 6). Over-activation will move the fulcrum points, around which the teeth are tipping, closer to the apices and increase the retroclination produced during overjet reduction. Activation may be checked by placing the appliance in the mouth with the cribs lightly over the first permanent molars. The passive retractor should lie just behind the incisal edges.

Bite plane trimming

In order to prevent re-eruption of the lower incisors, it is necessary to maintain a bite plane on the appliance for as long as possible during overjet reduction. Space is created by gradually trimming away acrylic behind the incisors according to the contour shown in figure 7. It is particularly important to create adequate space at the gingival margin or else the gingiva may pile up and hinder tooth movement (fig. 8). During the final 1–2 mm of overjet reduction the bite plane may be completely ground away from the appliance and the canine stops cut off.

Alternative method for overjet reduction

When the overjet increase is considerable, the Roberts' retractor is the spring of choice. When less than 4 mm of movement is needed, however, the self-straightening spring or strap spring is an acceptable alternative (fig. 9). Although less efficient than the Roberts' retractor, the strap spring appliance is well tolerated and less likely to break.

The strap spring is sometimes criticised for flattening the front of the arch. This effect is prevented by using a spring of adequate length. The method by which tapes welded to the arch are used to attach the ends of the spring ensures that the retractor is free-running and properly supported (fig. 10).

Stage 4. Retention

The importance of achieving full correction of increased overbite and overjet has been stressed repeatedly. Once a result has been obtained, an adequate period of retention is necessary, in order to allow time for the function of the lips and tongue to adapt to the new tooth position, and for the periodontal tissues to repair the changes induced during tooth movement. Every orthodontic result will relapse to some degree, but the extent of relapse is minimised by retention of sufficient firmness and duration. Six months of full-time retainer wear followed by 2 months at night only is an absolute minimum when an overjet has been reduced.

It is difficult to adjust an active retraction appliance so that it can be used as a passive retainer and a separate retention appliance is normally advisable. The standard Hawley retainer with a 0·7 mm short labial bow is ideal. U-loops must be incorporated to permit some adjustment. When it is wished to retain incisor rotations that have been induced during overjet reduction a fitted bow is best, otherwise there is little to choose between a fitted and a plain bow.

Summary

Proper attention to canine correction and bite opening is the foundation of successful overjet reduction, this phase of treatment must not be hurried.

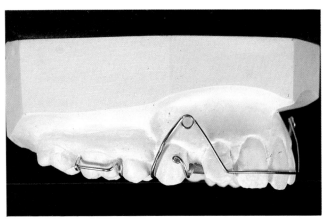

Fig. 4   A Roberts' retractor.

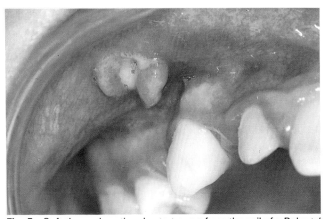

Fig. 5   Soft tissue ulceration due to trauma from the coil of a Roberts' retractor. Reproduced by kind permission of Wolfe Medical Publishing.

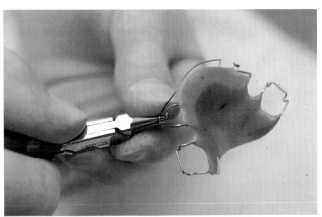

Fig. 6   Activation of the Roberts' retractor.

Fig. 7   Correct trimming of a flat bite plan to allow overjet reduction.

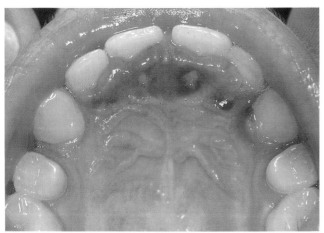

Fig. 8 Gingival hypertrophy due to incorrect bite plate trimming and possibly too rapid overjet reduction. Reproduced by kind permission of Wolfe Medical Publishing.

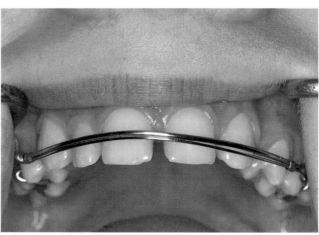

Fig. 9 A strap spring.

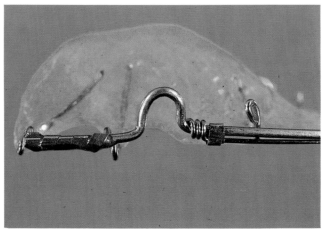

Fig. 10 Use of narrow tape welded around the buccal archwire as the tether for a strap spring.

A correct Class I canine relationship must be achieved before overjet reduction is begun. Overbite correction will often be the most difficult tooth movement to achieve, yet it is essential if the result is to be good in terms of aesthetics and stability.

When overbite is increased at the beginning of treatment, each appliance must carry a flat anterior bite plane. The bite plane must be built up as treatment progresses and trimmed away gradually during overjet reduction.

References

1   See chapter 23.
2   Foster T D. *A textbook of orthodontics.* 2nd ed. pp 36–37. Oxford: Blackwell Scientific Publications, 1982.
3   Adams C P. *The design, construction and use of removable orthodontic appliances.* 5th ed. pp 175–178. Bristol: Wright, 1984.
4   Houston W B J. *Walther's orthodontic notes.* 4th ed. p 83. Bristol: Wright, 1983.

25

Functional Appliance Therapy

W. P. Rock

Functional appliance therapy can give good correction of Class II division 1 malocclusion. The keys to success are careful case selection and use of appliances that are active enough to work, but comfortable enough to be worn for 14 hours each day. Patient motivation is also an important factor.

The first functional appliance of real success was the Andresen appliance. Originally described in 1936, this appliance became well established in Europe, especially for the treatment of Class II division 1 malocclusion. Since that time, other functional appliances have been devised, some as developments of the Andresen appliance, others according to new concepts. The interest that any technique holds for clinicians depends very much upon the degree of success that it achieves and with functional appliances there are no half measures. In any situation in which a functional appliance is used, it will either work very well or hardly at all. Success follows good case selection, favourable growth of the face and jaws, and prolonged and consistent appliance wear by the patient.

It will be apparent that none of these factors can be predicted with complete accuracy, and this could be taken as discouragement for any use of a functional appliance. However, in a proportion of cases, functional therapy provides spectacular and rapid occlusal correction without the need to extract teeth. These are the results that must be set against those cases in which a functional appliance finds its way into the model box soon after being made. It cannot be stressed too strongly that, whilst the practical management of functional appliances is relatively straightforward, case selection requires skill and perception. This is the challenge!

Mode of action

Current interest in functional appliances is centred upon the concept of treating malocclusion to profile-orientated goals. Functional appliances work best in the correction of mandibular retrusion associated with Class II malocclusion. When mechanical appliances are used to reduce the overjet in such a malocclusion, the extraction of four premolars is often an essential part of the treatment plan and this approach may have an undesirable effect on facial contour. Functional therapy, on the other hand, achieves its effect by encouraging forward movement of the mandibular dentition, leaving the maxillary teeth relatively unaffected. The result is correction of the unsightly retrusion of the lower arch that was the underlying problem. With this correction comes improved profile and a pleasing result. Perhaps the best way to describe the action of functional appliances is to say that they are inefficient at moving individual teeth but highly effective in altering the relationship of the maxillary and mandibular arches.

It is difficult to be certain how functional appliances achieve their effects since they are used in the growing patient, where even the use of serial lateral skull radiographs cannot demonstrate fully those changes that might have occurred without treatment. At the present time it has not been demonstrated conclusively that functional appliances influence growth of the mandibular condyle, and it may be that occlusal correction is brought about largely by tooth movement, followed by remodelling of the alveolar bone.

Types of functional appliances

There are two main subdivisions:

(1) Monobloc appliances that fit between and against the teeth. These consist of a large central acrylic portion carrying a labial bow, for example the Andresen appliance (1936), and the Harvold activator (1974).

(2) Vestibular appliances that fit around and about the teeth, providing screening labially and buccally against muscle pressure, for example the oral screen (1912) and the Frankel appliance (1969).

Irrespective of individual shapes, all functional appliances are loose interdental splints that harness, or otherwise control, the action of the muscles of mastication and facial expression. They all work best in the Class II division 1 malocclusion and it is for this application that they will be considered.

Although there are claims and counter-claims for the merits and disadvantages of different functional appliances, there is little to choose between the different designs in many cases. It is much better for the general practitioner to attempt to become familiar with one type than to indulge in constant experimentation. The appliances used most are the original Andresen activator, the Harvold activator, and the Frankel regulator (FRI) (fig. 1).

The Frankel has become something of a vogue appliance and there is no doubt that, in the right case, it confers advantages that the monoblocs do not. In particular, Frankel therapy may produce an increase in the size of the dental arches. Use of the appliance can resolve crossbites and also create space to relieve mild crowding of incisors or premolars (fig. 2). However, the Frankel is a complicated appliance and construction is time-consuming, as reflected by the high fees charged by commercial laboratories. The complex structure is vulnerable to damage and when this occurs it is often impossible to restore original form so that a completely new appliance must be made. A third disadvantage is that there is a tendency for the Frankel appliance to procline the lower incisors badly unless it is used with care. For all of these reasons, the general dental practitioner would be wise to choose a monobloc appliance when a functional approach is indicated. The discussion that follows will concentrate upon the Harvold activator. This requires less trimming than the original Andresen monobloc and is somewhat simpler to fit (fig. 3).

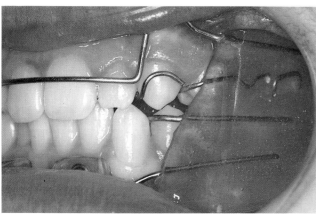

Fig. 1   The Frankel appliance.

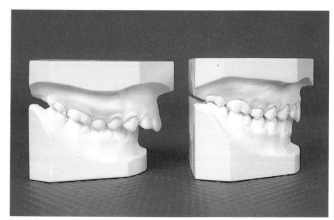

Fig. 2   Before and after models of a case treated using a Frankel appliance.

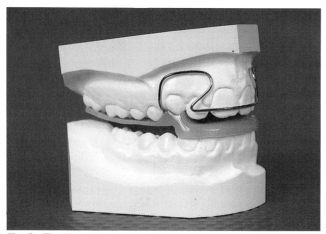

Fig. 3   The Harvold activator.

Case selection

Removable, fixed and functional appliances must not be regarded as separate entities, but rather as complementary techniques, used in combination to treat malocclusion. With this in mind, functional appliances are used in two main ways:
(1) To achieve complete correction in a carefully selected group of mild Class II division 1 cases which demonstrate certain well defined clinical features.
(2) To provide partial correction of the arch relationship in a wide range of more severe Class II malocclusions. Following this phase of treatment, detailed alignment is carried out by fixed appliances.

It is best for the general dental practitioner to confine the use of functional therapy to cases of the first type, carefully selected according to the following criteria:
(a) Uncrowded arches. As stated earlier, monobloc appliances do not affect individual tooth alignment.
(b) Molar relationship not exceeding one half unit Class II.
(c) Mild Class II skeletal discrepancy in both antero-posterior and vertical directions. A guide to antero-posterior discrepancy can be obtained by palpating the A and B points as described in chapter 23. With the Frankfort plane horizontal, Point B should be around 2 mm behind Point A. Differences of up to about 6 mm may be corrected completely by functional therapy.

Vertical skeletal discrepancy is also important, since it is likely to reflect the principal direction of mandibular growth. The mandible grows downwards and forwards away from the skull base and cases selected for functional therapy should have a predisposition towards forward growth. This is indicated by a face of normal proportions, a gonial angle that is square rather than obtuse and an overbite that is within normal limits. A long face, with high gonial angle and open bite, indicates a predominantly downwards direction of growth that may not respond well to functional therapy.
(d) Patient age. The effectiveness of functional therapy depends greatly upon favourable growth during the period of appliance wear. Growth is not constant but periodic.[1] The beginning of the main growth spurt usually occurs between 12 and 14 years, being somewhat earlier for girls than for boys. In order to obtain a really accurate assessment of the growth pattern of an individual, it is necessary to compare serial height measurements against standard growth and development records. However, a reasonable idea can be gained by measuring a child at intervals of 3 months and beginning treatment when growth suddenly appears to accelerate.

Clinical procedures

These include:
(1) Prefunctional treatment with removable appliances.
(2) Impression taking for functional appliances.
(3) Bite recording.
(4) Fitting the activator and instructions to the patient.
(5) Review appointments.

Prefunctional treatment

As stated previously, a suitable case should have well aligned arches and a molar relationship not exceeding one unit Class II. The adequacy of final occlusion may be assessed by placing study models into a Class I molar occlusion. This should produce a normal overjet. A prefunctional phase of treatment may be advisable in one or both of the following circumstances:
(a) Proclination and spacing of the incisors.
(b) Buccal segment crossbite produced by forward posture of the mandible. A small amount of expansion provides a good post-treatment occlusion.

These corrections are simultaneously achieved by an expansion and labial segment retraction appliance (ELSRA), (fig. 4). The labial bow is automatically activated as the mid-line screw is turned. Acrylic must be trimmed away

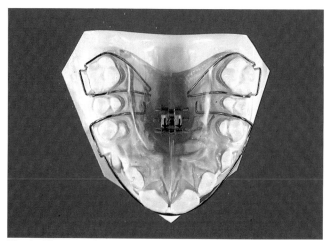

Fig. 4 The expansion and labial segment retraction appliance.

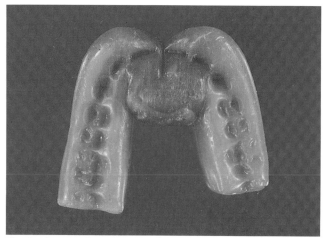

Fig. 5 A wax roll used to record the construction bite for a Harvold activator.

behind the incisors in order to provide space into which they can move.

Even when forward positioning of the mandible does not produce a crossbite, it may still be desirable to attempt to retrocline proclined incisors and close spaces. These movements may be achieved using any standard appliance for overjet reduction, such as a Roberts' retractor.

Another useful gain from the prefunctional phase is that it enables the monobloc to be fitted without trimming acrylic away from behind the maxillary incisors. It is easier for a patient to come to terms with a close fitting, untrimmed monobloc than with a very loosely fitting one.

Impression taking
Well extended, muscle trimmed impressions are essential for all functional appliances. In the case of the activator, the most critical area is the lingual sulcus. The lingual flange of the appliance must be fully extended in this area to provide stability, but not over-extended so that it produces ulceration.

Bite recording
Many theories have been expressed to explain the mode of action of the activator, but none is wholly convincing. However, there is no doubt that the moving force of the appliance is generated by muscle stretching produced by the fact that the appliance is constructed to a postured bite. In practice, two somewhat conflicting considerations must be borne in mind when recording the bite:
(1) The sagittal and vertical displacement of the mandible must be sufficient to stretch the muscles and activate the appliance.
(2) The size and shape of the appliance must be such that it can be worn consistently for at least 14 hours each day.

The mandibular protrusion should be straight forwards, with no lateral deviation. If case selection has been rigorous, there should be no local tooth irregularities and a straight protrusion is indicated by coincidence of upper and lower incisor centre lines. For a mild to moderate Class II arch relationship the correct construction bite will be such that the incisors are edge to edge sagitally and 5 mm apart. Before taking a bite registration it is vital to spend time instructing the patient in what is required. The patient should be seated in an upright, but relaxed position and gently shown how to

achieve the desired bite. If an edge to edge incisor bite is used, the thumb nail of the operator may be used as a convenient guide. Only when the child can bite reliably and repeatedly into the position should the wax bite wafer be introduced.

A wax roll of finger thickness is prepared from a well softened sheet of pink wax. The sheet should be folded or rolled gradually and returned to the flame so that hard spots are eliminated. These tend to deflect the mandible and spoil the recording.

The softened roll is formed into the shape of the dental arches and identified anteriorly so that, when it fits over the teeth, the incisors are left clear (fig. 5). This precaution allows the operator to check for mandibular deviation during bite recording. Some orthodontists incorporate a wooden tongue spatula into the wax bite to act as a handle. With the wax in place over the lower teeth, the patient is asked to bite slowly into the previously determined posture. When the desired position is achieved, the wax is removed from the mouth and chilled. It is then reseated in the mouth to check that no distortion was produced during removal from the teeth.

Sometimes the presence of bulky wax in the mouth will cause the child to bite in a manner far removed from the rehearsed position. It is important that the operator should not display irritation if this occurs. The child–orthodontist relationship is of paramount importance in functional therapy, since some children find difficulty in accommodating to the bulky appliances. Such difficulties are minimised if the orthodontist has the trust and confidence of the child. After a period of further training, the bite recording is repeated. An excellent account of bite recording for functional appliance construction is given by Gaber and Neumann.[2]

Fitting the appliance and instructions to the patient
Little or no trimming should be needed if the activator is properly made following accurate impression taking. However, the lingual flanges may require reduction, particularly over the area of a prominent mandibular torus. When the activator is in place, a check should be made for coincidence of the centre lines (fig. 6). The heavy lower incisor capping shown is necessary in order to prevent incisor proclination. This is a crucial feature of design. If the lower incisors are not properly capped, the activator should not

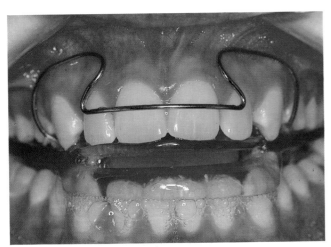

Fig. 6 An activator in the mouth

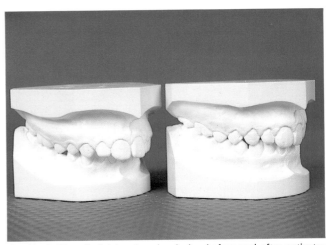

Fig. 7 A Class II division 1 malocclusion before and after activator treatment.

be fitted. The bow should be activated only lightly.

An activator is the largest device that any branch of dentistry expects a patient to wear for a long period and proper instruction and motivation of the patient is vital if therapy is to succeed. Parents will also express surprise, even dismay, at the size of the appliance and the strange appearance and speech of the child when it is in place. Such natural reactions should be countered positively, with explanations of the benefits and efficacy of the appliance: extractions are not required, the activator need not be worn to school, an excellent final bite will be achieved.

Some children will happily wear the Harvold full time. When this is the case, results are both rapid and successful. However, some children will not cope with such a demand and part-time wear is the best that can be achieved. The least commitment that will produce success is around 14 hours of wear per day. The appliance should be inserted after the evening meal, so that it is worn for a few hours before the child goes to bed. This precaution minimises the likelihood of the appliance being rejected during the night. Some orthodontists counter this tendency by incorporating into the activator Adams' clasps around the maxillary first molars. The added retention gives confidence and security to the child during the first few weeks of treatment. Once the patient can sleep with the appliance in place, the clasps are clipped off.

Review appointments
The patient should be seen 2 weeks after appliance fitting in order to assess wear and to provide encouragement. High

spots on the acrylic may also be eased at this time. Provided that all is well, further recalls may be at intervals of 2 months.

Little change in the occlusion is to be expected for 3 months or so. During this time the patient must be encouraged and reassured that the activator is working correctly. The first indication of real progress comes when the patient is unsure of where to bite when asked to do so. Once again the enthusiasm of the operator must be transmitted to the child. Complete occlusal correction usually takes around 18 months. When this has been achieved the now passive activator may be worn as a retainer for a further 6 to 9 months. Providing that the occlusion has good intercuspation, relapse is less likely after this time (fig. 7).

Summary
Successful resolution of a Class II division 1 malocclusion by functional therapy involves good case selection, favourable growth and dedicated appliance wear.

The Harvold activator is a simple and robust functional appliance that works well in the right case; a Class II arch discrepancy is corrected without the need to extract teeth. A prefunctional phase of treatment, using removable appliances, is beneficial in some situations.

References
1 Tanner J M, Whitehouse R H. *In* Braham R L and Morris M E (eds). *Textbook of pediatric dentistry.* 2nd ed. pp 5–7. Baltimore: Williams and Wilkins, 1985.
2 Graber T M, Neumann B. *Removable orthodontic appliances.* 2nd ed. pp 175–197. Eastbourne: W B Saunders Company, 1984.

26

Treatment of Class II Division 2 Malocclusions

W. P. Rock

Severe Class II division 2 is a difficult malocclusion to treat properly, especially when lower incisors contact palatal mucosa. Only fixed appliance therapy can achieve worthwhile and lasting improvement of the incisor relationship in this situation. In less severe cases, it may be possible to achieve limited objectives using removable appliances. When planning treatment, extraction of mandibular premolars should be avoided if at all possible.

Treatment planning for Class II division 2 malocclusions is dominated by the need to correct the increased overbite that is a constant feature in this type of case. An earlier chapter described how overbite stability is dependent upon two factors:
(1) inter-incisal angle;
(2) biting position of the lower incisors.

In the Class II division 2 malocclusion, both upper and lower incisors are retroclined by the action of a tight lower lip so that the inter-incisal angle is high. A situation is therefore created in which incisor over-eruption and overbite increase are likely. These possibilities are made almost certain by the vertical relationship of the jaws in a typical Class II division 2 patient (fig. 1.). The gonial angle, that is the angle between the ramus and body of the mandible, tends to be square, so that the maxillary–mandibular plane angle (MMPA) is less than the average of 28° and lower face height is reduced. It is the square jaw and high cheekbones that produce the characteristic facial appearance so often associated with the malocclusion. Reduced lower face height naturally favours the production of overbite increase, which may result in trauma:
(1) by the action of the lower incisal edges on the palatal mucosa behind the maxillary teeth;
(2) by the action of the upper incisal edges on the mucosa of the alveolus labial to the lower incisors.

The full consequences of a traumatic incisor bite may not be apparent in the child, possibly due to the fact that over-eruption of the lower incisors is cancelled out by continuing facial growth. However, in middle life, when facial growth has ceased, severe irreversible periodontal damage may be produced by incisors that bite against the mucosa of the opposing jaw (fig. 2). This presents an intractable problem that may require the wearing of a permanent bite guard to protect the teeth and gums from further damage.

The question that governs treatment planning in Class II division 2 malocclusions is therefore: 'Is the overbite traumatic, or will it become traumatic in later life if left untreated?' If either the tips of the lower incisors, or the tips of the upper incisors, or both, contact mucosa, it is wise to assume that the answer to the question is 'Yes', and to plan treatment accordingly. Worthwhile permanent correction will only be achieved by fixed appliances that can:
(1) reduce overbite efficiently to establish a cingulum biting position for the lower incisors;
(2) reduce inter-incisal angle by altering the axial inclinations of the incisors during movement;
(3) reposition the lower incisors accurately according to

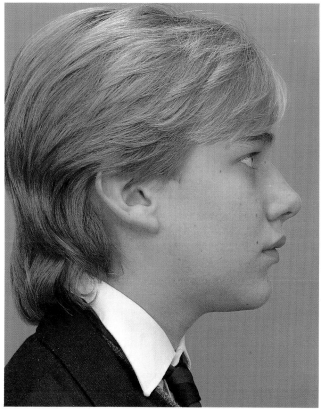

Fig. 1 A typical Class II division 2 face.

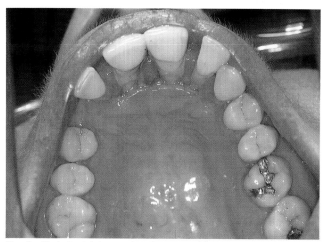

Fig. 2 Periodontal destruction due to overbite trauma.

profile-orientated planning criteria.

Fixed appliances are much more effective than removables at producing bite opening, since the curve of Spee is flattened by forces generated in the archwire. A removable appliance simply prevents eruption of the lower incisors by causing them to bite against a flat anterior bite plane while the posterior teeth continue to erupt. With either system, effective bite opening relies upon favourable jaw growth and tooth eruption, so that fixed appliance treatment attempted after an unsuccessful period of management with removable therapy may be too late to achieve optimal effect.

The ability of any appliance to open the bite is also hampered by extraction of teeth. British orthodontic textbooks have stressed the need for extractions when treating the Class II division 2 malocclusion.[1-3] This situation was confirmed for a time by the popularity of the Begg technique, in which premolar extractions were usually essential.[4] By contrast, American texts have suggested that extraction in Class II division 2 be practised only where there is definite basal arch discrepancy and not as an aid to tooth alignment.[5] This latter approach is of particular importance in Class II division 2 malocclusions, in which the deep overbite leads so often to soft tissue trauma. Bite opening by fixed appliances is most effective when all premolars and molars are present and in contact. There is a real risk that if mandibular premolars are extracted, as part of an unsuccessful attempt to treat the case with removable appliances, subsequent fixed appliance treatment may become much more difficult, if not impossible.

In summary, when planning treatment for a Class II division 2 malocclusion, a basic decision must be made at the outset:

(1) To use removable appliances and accept the overbite and the positions of upper and lower central incisors and extract teeth so that space is created for alignment of the upper lateral incisors and any other misaligned teeth.

(2) To embark upon a treatment plan involving fixing appliances. This will have several advantages:

(a) reduction of the overbite to give a cingulum biting position for the lower incisors.

(b) repositioning lower incisors to the A-Po plane.

(c) alteration of the inter-incisal angle by torquing the roots of both upper and lower incisors to ensure that the bite reduction will be permanent. Such a result can be visualised on a cephalometric tracing of the type sometimes known as a 'visual treatment objective' (fig. 3).

Treatment planning for Class II division 2 with removable appliances

Since the position of the upper central incisors and the overbite are to be accepted, the principle treatment objective will be to align the upper lateral incisors from their characteristic crowded and labially rotated positions. The amount of space needed to complete alignment of the lateral incisors will depend upon the extent of the initial displacement, but in most cases it will be around half a unit of premolar space, that is 4 mm. This space may be created in one of two ways:

(1) by retraction of the buccal segments;

(2) by extraction of premolars.

Buccal segment retraction

This is an attractive treatment approach when up to half a unit of space is needed, since it avoids leaving residual space

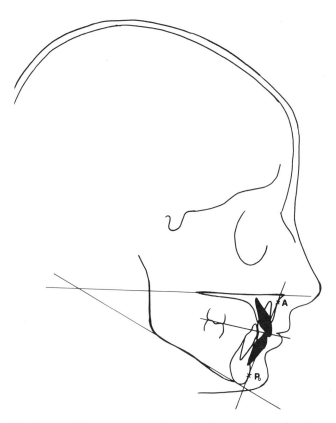

Fig. 3   A cephalometric tracing showing incisor positions before *(clear)* and after *(solid)* treatment with fixed appliances. The tips of the lower incisors have been moved to the A-Po line.

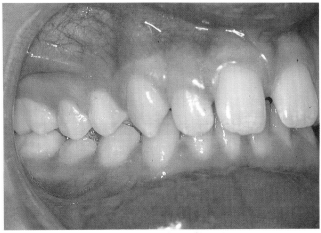

Fig. 4   An occlusion that is one half-unit Class II.

in the premolar region. It is indicated particularly when the first permanent molars occlude half a unit Class II so that they are cusp to cusp with the lower teeth (fig. 4). Successful distalisation of the upper molars will relocate them into a Class I bite.

Active buccal segment retraction requires the use of headgear. Even then, the molar stacking that is often seen on radiographs in this malocclusion may prevent successful retraction of the first permanent molars unless the second molars are first extracted (fig. 5). The decision to remove maxillary second molars may naturally lead on to compensatory extraction of the mandibular second molars. This may be desirable for the following reasons:

(1) Following extraction of the maxillary second molars, the unopposed mandibular teeth may over-erupt and traumatise

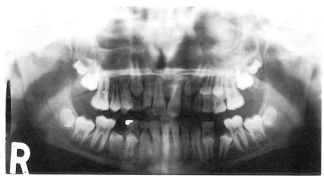

Fig. 5 A panoramic radiograph showing stacking in the maxilla.

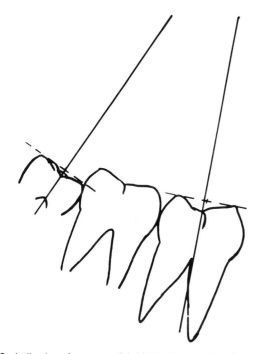

Fig. 6 Indications for successful third molar eruption after extraction of mandibular second molars.

the mucosa of the maxillary tuberosity.

(2) Lower second molar extraction may successfully relieve mesio-angular impaction of the third molars and prevent the need for surgery. However, there is also a risk that removal of the lower second molar may convert a mesio-angular impaction of the unerupted tooth into a more serious horizontal impaction. This unfortunate complication is unlikely to occur if the following conditions are satisfied (fig. 6):[6]

(1) Long axis of third molar at not more than 30° to that of the first molar.

(2) Contact point of third molar at the level of, or above, amelocemental junction of second molar.

(3) Age 12–15 years; bifurcation of third molar beginning to calcify.

Appliance designs for molar distalisation

Successful molar distalisation requires the use of extra-oral forces in the region of 400–500 g per side. It is difficult to resist the tendency of such high forces to displace a removable

appliance and many orthodontists prefer to fit bands to the anchor molars and to use a clip-over type of removable appliance (fig. 7). The widespread use of preformed bands with prewelded attachments has brought this method within the reach of the general dental practitioner, especially since several orthodontic suppliers will provide a 'starter' band kit incorporating a range of popular sizes. Since there are obvious dangers concerning the application of high extra-oral forces to banded teeth, appropriate postgraduate instruction should be obtained before embarking upon such treatment.

When only a limited amount of distalisation is needed, the design shown in figure 8 may be successful. Since lower extra-oral forces must be used when the anchor teeth are not banded, it may be wise to turn one screw at a time in order to minimise the risk of anchorage slip. Retraction is continued until Class I buccal segment occlusion has been achieved.

Premolar extraction

Buccal segment retraction is not indicated if the first permanent molar occlusion is already Class I. Instead, space for incisor alignment is created by premolar extractions.

Extraction of mandibular teeth

The previously held theory that the lower incisors would somehow 'collapse' lingually following premolar extractions was largely disproved by the classic work of Mills.[7] However, although the lower labial segment is remarkably stable, it remains unwise to extract mandibular premolars from a Class II division 2 malocclusion where the incisal overbite is just acceptable on the gingival third of the palatal surface of the maxillary incisors. In this situation, the least amount of lingual movement by lower incisors may convert a tenuous overbite into one that is traumatic in the gingival crevice of the maxillary teeth. For this reason it is better to accept mild lower labial segment crowding in this malocclusion than to create excess mandibular space with possibly disastrous effects upon overbite stability.

In general, it is wise to avoid extraction of mandibular premolars when removable appliances are to be used to treat Class II division 2 malocclusions.

Extractions in the upper jaw

The extractions of choice must be the first premolars, unless special circumstances, such as palatal displacement or hypoplasia, point to the second premolars. Second premolar extraction often results in a poor buccal occlusion and unsatisfactory contact between first premolar and first molar. It used to be said that one advantage of second premolar extraction was that any residual space left at the end of treatment would be further back in the mouth and therefore less noticeable than residual first premolar space. Such a claim is an admission of the acceptance of poor quality treatment results and can no longer be defended.

Unfortunately, it is likely that residual spacing will be left following the extraction of first premolars. This situation arises owing to the special problems associated with canine relationship in the Class II division 2 malocclusion.

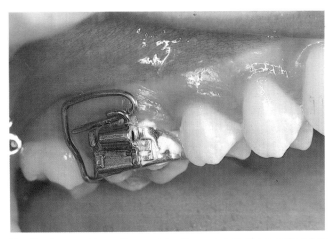

Fig. 7 A removable appliance retained by means of clips over the tubes on molar bands.

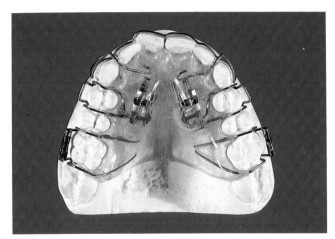

Fig. 8 A twin-screw appliance for buccal segment retraction.

The special problem of canine relationship in Class II division 2 malocclusions

In an early chapter in this series, the importance of achieving a correct Class I canine relationship during treatment was stressed. When the upper and lower canines occlude correctly in a Class II division 1 situation, the proper amount of space for overjet reduction is available, provided that overbite is normal. In the division 2, however, the situation is complicated by retroclination of both sets of incisors. This effect produces a 'hidden' overjet that cannot be corrected. It follows, therefore, that if the upper canines are retracted fully into a Class I relationship with the lowers, more space will be created than is needed for alignment of the upper incisors and spacing will result. It is possible to avoid residual spacing mesial to the canines by leaving these teeth in a half-unit Class II relationship with the lowers, but this decision produces another unfortunate consequence. The upper canines are much wider bucco-palatally than incisors and they appear somewhat prominent in any degree of Class II relationship.

The mechanics of canine retraction are no different from those in any other malocclusion, except that the canines in the Class II division 2 are often mesio-labially rotated and buccal canine retractors may be preferable to palatal ones. A flat anterior bite plane is unlikely to produce a permanent bite reduction but any help in this direction will be beneficial and is worth a try.

Correction of the upper lateral incisors

These teeth tend to be displaced in two ways:
(1) they are mesially inclined;
(2) they are mesio-labially rotated.

These displacements are inter-related and complete correction requires that the apices of the lateral incisors are moved mesially at the same time as the rotation is corrected. Relapse is very likely if the lateral incisors are simply pushed back into the line of the arch without any attempt being made to correct mesial angulation.

The most satisfactory way to simultaneously upright and rotate an incisor is to use a whip spring (fig. 9). The spring is made from 0·016 inch (0·4 mm) Australian wire or similar. The annealed end is bent firmly under the wings of an edgewise bonded bracket or secured through the channels of a ripple bracket. By using a slightly different design of spring, a Begg bracket may also be used.[8] When passive the spring is

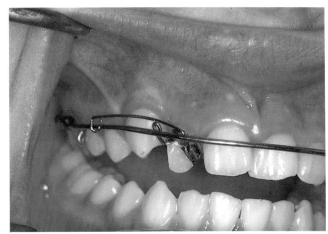

Fig. 9 A whip spring appliance.

bent to lie above and outside the line of the buccal arch. When pulled down and hooked under the archwire, it generates both uprighting and rotating forces. The whip spring is very efficient and must be activated only gently. Once the tooth is in the line of the arch it is wise to achieve a small degree of over-rotation by cranking the arm of the spring just mesial to the bracket. It is well known that corrected rotations are notoriously likely to relapse, although use of the whip spring does minimise this possibility by the uprighting that it produces. A minimum of 6 months' retention is needed, either by a fitted labial bow on a removable retainer, or by the use of a fixed retainer bonded to the palatal surfaces of the derotated tooth and its neighbours. Multistrand wire is suitable for this purpose.

Less severe lateral incisor displacement may be dealt with by means of a buccally approaching spring of the type shown in figure 10. This is suitable for tucking in the mesial corner of the displaced tooth. Since it is not possible to achieve uprighting and over-correction with this spring, relapse may be more likely. Prolonged retention is the best way to minimise this possibility.

Summary

Treatment of the Class II division 2 malocclusion with removable appliances must have limited objectives. The position of the maxillary central incisors and the overbite must be accepted and tooth movement limited to that required to create space for alignment of the upper lateral

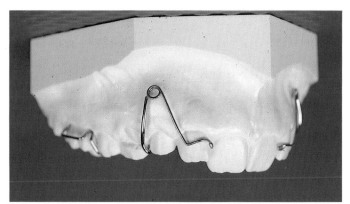

Fig. 10 A buccally approaching spring for lateral incisor alignment. This spring is made from 0·5mm wire supported to the coils by soft tubing.

incisors. Lower premolar extractions may compromise the stability of the overbite and should be avoided, even if this means that some lower incisor crowding must be accepted.

If either upper or lower incisal edges contact mucosa, removable appliance treatment involving extraction of teeth should not be attempted.

References

1 Hovell J H. *In:* Walter DP (ed). *Current orthodontics*. pp 250–252. Bristol: John Wright and Son, 1966.

2 Mills J R E. *Principles and practice of orthodontics*. pp 158–159. Edinburgh: Churchill Livingstone, 1982.

3 Houston W B J, Tulley W J. *A textbook of orthodontics*. Bristol: John Wright, 1986.

4 Begg P R, Kesling P C. *Begg orthodontic theory and technique*. pp 75–79. Philadelphia: W B Saunders, 1977.

5 Salzmann J A. *Practice of orthodontics*. Vol Two. p709. Philadelphia: J B Lippincott Company, 1966.

6 Cryer B S. Third molar eruption and the effect of extraction of adjacent teeth. *Dent Pract* 1967; **17:** 405–418.

7 Mills J R E. The stability of the lower labial segment. A cephalometric survey. *Dent Pract* 1968; **18:** 293–306.

8 Houston W B J, Isaacson K. Orthodontic treatment with removable appliances. p128. Bristol: John Wright and Sons, 1980.

27

Extra-oral Traction

W. P. Rock

Extra-oral traction has two main applications in orthodontic treatment. It may be used to prevent forward movement of anchor teeth and also to provide a force for distalisation of molars, and/or buccal segments. If correctly applied, EOT can help to ease problems in a difficult treatment and make possible an otherwise impossible treatment plan.

What is anchorage?

The forces used in orthodontics are governed by Newton's third law, 'To every action there is an equal and opposite reaction'. Whenever force is used to retract teeth, the reaction force has the ability, unless adequately resisted, to move other teeth forwards. Resistance to unwanted reaction forces comes under the umbrella of 'anchorage', which may be defined as 'the site from which a force acts'. There are three main types of anchorage: intra-maxillary, inter-maxillary and extra-oral. The first type is typified by an upper removable appliance designed to retract canines after removal of first premolars. Three tooth movements will tend to close the extraction spaces. The first is distal movement of the canines in response to spring force. The second is forward movement of buccal teeth in response to physiological mesial drift and the third is the reaction to the moving force. Unwanted forward movement of teeth, called anchorage slip, is demonstrated most clearly by overjet increase. It requires an effort of imagination to accept that an orthodontic appliance can actually move forwards in the mouth, but this is what must happen when anchorage is slipping.

Inter-maxillary anchorage is the principal method by which fixed appliance systems correct antero-posterior arch discrepancy. The upper and lower arches are joined by elastics, so that teeth in one arch provide anchorage for the movement of teeth in the opposing arch.

Extra-oral anchorage comes from outside the mouth. The site from which the force acts is the back of the patient's head. Extra-oral anchorage confers a great advantage over the two intra-oral types; high forces can be applied to blocks of teeth without a reaction being generated elsewhere in the mouth.

Extra-oral anchorage (EOA), extra-oral traction (EOT)

EOA is used to indicate a wish to maintain the existing positions of buccal segment teeth, whilst EOT implies active distal movement.[1] This distinction is somewhat arbitrary when considering the practicalities of extra-oral force application, since a degree of elasticity is needed to create a wearable system. However, it is a useful way of defining the objectives of extra-oral force application in a particular case.

Can EOT be avoided?

It is a considerable imposition for a patient to wear headgear and common sense suggests that it should be avoided whenever possible. In a marginal situation the need for EOT may be avoided by careful design and activation of a removable appliance:

(a) Careful design can maximise the anchorage value of an appliance. For instance the addition of a close fitting labial bow will reduce the chance of incisor proclination and thereby improve anchorage (fig. 1).

(b) Careful spring activation can minimise the strain on anchor teeth.

Anchorage slip

The unwanted tooth movements that constitute anchorage slip may be seen either when too many teeth are moved at once, or when too much force is generated by the appliance. An excessive retraction force may slow down or completely stop tooth movement. However, the reaction to such excessive force, when spread over anchor teeth, may be of the correct magnitude to produce anchorage slip. The two main indications that slip is taking place are:

(a) Overjet increase due to incisor proclination during the retraction of canines or buccal segment teeth.

(b) Alteration in first molar occlusion.

Closure of extraction spaces is not a reliable sign that anchorage is secure, since forward movement of buccal segment teeth may encroach upon the spaces to an equal, or greater, extent than retraction of anterior teeth. It is good practice to measure overjet at frequent intervals during treatment, in order to check that anchorage is not being lost.

When is EOT needed?

Although anchorage slip represents an obvious need for anchorage reinforcement, it is poor planning to have to introduce EOT in this situation. It is much better to include EOT as part of the initial treatment plan for those cases where anchorage is critical and make appropriate provision for it on every appliance used during treatment.

EOT is used in two main situations: to distalise buccal segments in an attempt to avoid premolar extractions and to

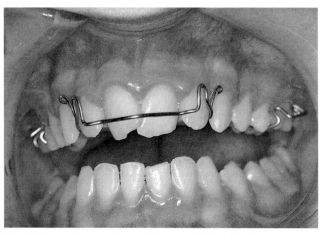

Fig. 1 A removable appliance with a close fitting labial bow.

stabilise or distalise maxillary first permanent molars as part of a treatment plan that also involves premolar extractions.

Distalisation of buccal segments

One way of avoiding residual space after orthodontic treatment is to make room for overjet reduction or relief of incisor crowding by moving back the buccal segments. This approach relies heavily upon dedicated headgear wear and even then it is unwise to be too ambitious. A half-unit Class II molar relationship presents few problems, but more severe discrepancy may require extraction of the maxillary second molars before retraction can succeed.[2]

EOT in association with premolar extractions

Extraction of first premolars is indicated when more than half a unit of space is needed to reduce an overjet or relieve incisor crowding. The following occlusal signs, alone or in combination, indicate that the removal of upper first premolars may not provide sufficient space for full occlusal correction:
(1) Molar relationship Class II;
(2) Canine relationship more than one unit Class II;
(3) Overjet exceeding 6 mm;
(4) Severe crowding of upper incisors;
(5) Lower incisor crowding.

Appliances for EOT

An EOT appliance has three parts:
(1) Headgear;
(2) Facebow, or other attachment for transmitting force from the headgear into the mouth;
(3) Intra-oral appliance.

Headgear

The effect of EOT depends upon the *direction, duration and amount* of the applied force.

Direction control depends upon suitable headgear design. Force should be applied along or above the occlusal plane. The two most frequently used directions of EOT are high-pull and regular-pull (fig. 2). Low-pull, as applied from a cervical strap (fig. 3), is now not recommended, since forces below the occlusal plane will displace a removable appliance that is retained by Adams clasps, if that force is of sufficient magnitude to be effective in anchorage reinforcement. A similar force applied to banded molars will have the more disastrous consequence of extruding the teeth. This problem will be considered later.

High-pull headgear

This type of headgear fits over the occiput (fig. 4). Force is applied from the headgear via elastic bands and curved 'J-hooks' to attachments on the labial bow of a removable appliance or the archwire of a fixed appliance. When applied to a fixed appliance, high-pull EOT is a valuable aid to anchorage reinforcement and to bite opening.[3] The latter effect is not possible with a removable appliance since the teeth are not held with sufficient firmness. Also, the palatal acrylic must greatly reduce the effect of the upward force.

The most obvious advantage that high-pull EOT gives to a removable appliance is improved retention. However, there is a much better way of ensuring that EOT force will not

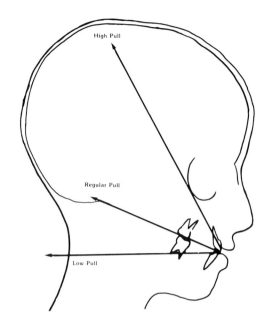

Fig. 2 The main types of EOT.

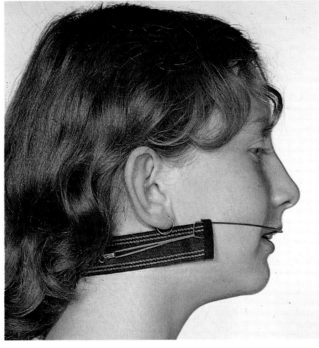

Fig. 3 A cervical strap.

displace a removable appliance; this is to band the anchor molars (see later).

Regular-pull headgear

Whatever direction of pull is used, a full headgear is recommended. Molar distalisation requires the application of 450 g (1 lb) to each side and a headgear is needed to provide comfort and security when such high forces are used.

Several unfortunate accidents have been reported, in which children have been injured by elastic recoil of a facebow. This risk can be eliminated by safety headgears which become detached from the facebow as a result of accidental or improper removal. However, some designs are not robust enough to provide good service over a prolonged period and a cheap and effective alternative is a simple safety strap which hooks over the ends of the facebow and prevents it

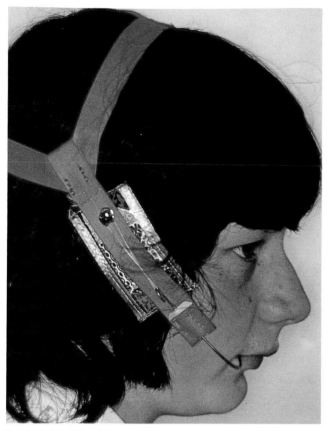

Fig. 4 High-pull headgear.

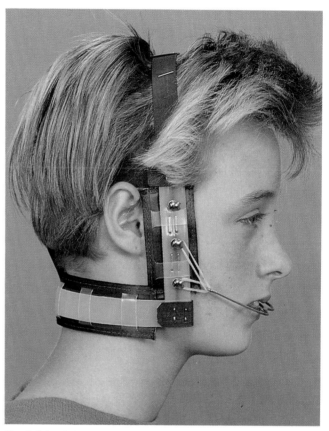

Fig. 5 A Lee Laboratories headgear.

from being pulled forwards.

The Lee Laboratories headgear* (fig. 5) is easily fitted and comfortable to wear. The three attachments allow for a direction of pull to be used that is slightly above the occlusal plane so that appliance displacement is minimised. The headgear is supplied flat and assembled using a stapler. It is important to ensure a comfortable fit around the ears.

Duration of wear must be adequate if EOT is to succeed. Headgear must be worn for 12–14 hours each day in order to achieve appreciable distalisation of teeth. For extra-oral anchorage it is usual to instruct the patient to wear the headgear for only 10–12 hours. It is wise not to confuse a patient by varying the instruction concerning duration of wear, but to insist that it is followed whenever headgear is worn. The effectiveness of EOT may be altered as required by using elastics of different strength or by withdrawing the headgear entirely.

Amount of force needed depends upon the intended function of the headgear. Active distalisation of molars requires 450 g (1 lb) to each side but 250 g (0·5 lb) is sufficient to stabilise the position of the anchor teeth and ensure that there is no slippage. Special elastics are available that deliver a designated force when stretched to twice their passive length (fig. 6). The distance between the facebow arm and the hook on the headgear is a guide to choice of correct elastics for a particular situation.

Facebow

Facebows, also known as Kloehn bows, are available from all major orthodontic suppliers, usually in a range of five sizes. The most important property that the bow must possess is

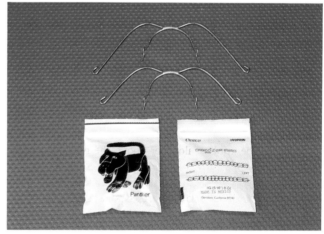

Fig. 6 Short and super-short facebows, with Zoo elastics.

rigidity, so that it will not bend when EOT forces are applied to it during treatment. When a full headgear is used, there is only a small amount of space for elastics and short outer arms must be used. Most facebows are made with long, short and super-short outer arms (fig. 6). The correct size will be one that leaves about an inch between the ends of the arms and the headgear hooks.

The length of the inner bow should be such that, when the U-loops contact the molar tubes, the bow should be just clear of the teeth. The bow is adjusted so that it lies between the lips when they are at rest (fig. 7).

Intra-oral appliance

The inner arms of the facebow fit into buccal tubes that may be soldered to the bridges of Adams' clasps (fig. 8), or welded to molar bands. Alternatively, with the En-masse appliance,

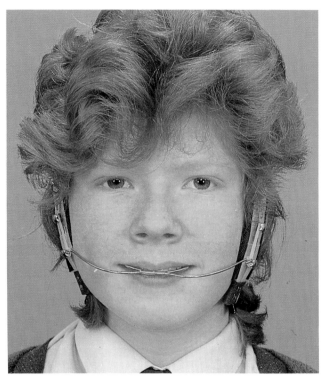

Fig. 7 A facebow correctly adjusted to lie between the lips.

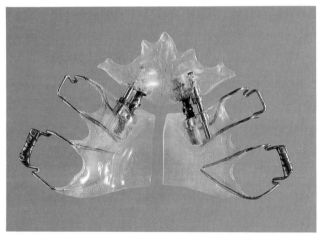

Fig. 8 EOT tubes soldered to the bridges of Adams' clasps.

the facebow may be attached directly to the appliance (fig. 9). The coffin spring is activated slightly to prevent the creation of cross-bites during molar distalisation.

The usefulness of EOT is limited when it is applied to a standard removable appliance. Clasps may be distorted during facebow fitting and retention of the appliance may be inadequate to resist displacement by heavy EOT forces. These disadvantages are overcome if bands are fitted to anchor molars. Preformed bands, already fitted with welded buccal tubes, are readily available and many orthodontic suppliers provide a starter band kit at modest cost. It is not possible to use Adams' clasps on banded molars; instead the removable appliance is retained by recurved clips that engage above the buccal tube on each band (fig. 10). The usual clinical procedure is first to fit the bands and then to take an impression for construction of the removable appliance. This may be designed so that EOT provides the only retraction force or screws may be incorporated to facilitate tooth movement. Screws are angled to act along the line of the teeth so that retraction does not produce cross-bites.

Beware! Incorrect application of EOT to anchor teeth that are banded will result not in displacement of the appliance, but of the teeth. Molar extrusion may produce an anterior open bite that can prove very difficult to resolve.

There have been several accidents in which patients have been injured by elastic recoil of a facebow. Care must be taken to ensure that the bow is securely seated in the buccal tubes before elastics are added and some form of safety device should be used. Elastics should be disconnected before the facebow is removed from the mouth.[3]

Summary

EOT is a valuable adjunct to orthodontic treatment. The additional anchorage provided may avoid the need to remove premolars in cases with mild crowding and in severely crowded cases it may avoid the need to extract more than two teeth in the upper arch.

EOT works best when the anchor teeth are banded. However, the direction of pull must then be carefully arranged to avoid unwanted displacement of the anchor molars.

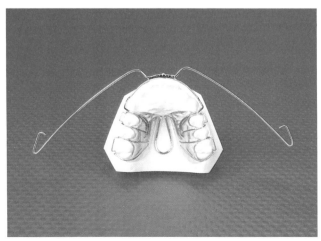

Fig. 9 A coffin spring made from 0·9 mm wire.

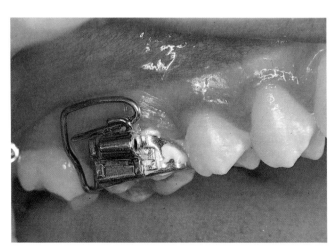

Fig. 10 A retaining clip fitting over the buccal tube on a molar band.

Headgear must be worn for more than 12 hours each day, with a force of 450 g on each side when teeth are to be distalised. Anchorage stabilisation requires rather less wear and a force of 250 g.

The efficacy of EOT may be judged by attainment and maintenance of the desired molar relationship.

References

1 Foster T D. *A textbook of orthodontics.* 2nd ed. pp 240–243. Oxford: Blackwell Scientific Publications, 1982.

2 See chapter 26.

3 Mills J R E. *Principles and practice of orthodontics.* pp 139–142. Edinburgh: Churchill Livingstone, 1982.

4 Houston W J B, Isaacson K G. *Orthodontic treatment with removable appliances.* 2nd ed. pp 44–48. Bristol: John Wright and Sons Ltd, 1980.

Lee Laboratories headgear: Orthomax Dental Ltd, Bankfield House, Carrbottom Road, Bradford BD5 9BJ.

Zoo elastics: Ormco Zoo Elastics, Oradent Ltd, 59 Eton Square, Eton, Windsor, Berkshire SL4 6BQ.

28

Essential Steps in the Training of Professionals

M. E. Pendlebury and G. Brown

Many general dental practitioners in the years ahead will be involved in training young dentists. A knowledge of training as well as a knowledge of dentistry is crucial for dental vocational training to be effective. Because a dentist knows how to remove an impacted wisdom tooth, it does not follow that he knows how to train someone else to do it. Training is a complex, challenging task which involves rather more than providing a surgery, occasional discussions with a trainee and a day-release scheme of lectures. Hence this chapter explores such essentials as the nature of training, the tasks of trainers, trainees and course organisers and the implications of these tasks for training dental vocational trainers.

Before beginning the discussion of training and the training course, it might be useful to provide a brief sketch of the development of dental vocational training. The first dental vocational training scheme was established as a pilot experiment based at the Guildford Postgraduate Medical Centre under the auspices of the Postgraduate Dental Dean of the British Postgraduate Medical Federation.[1] This provided a template for further schemes which were described at a workshop entitled 'Vocational training: the nuts and bolts' in 1982.[2] Trent region appointed a course organiser in 1979 and the first intake of trainees was in January 1981. Like most schemes, it now provides one year of training in which 30 days are spent in day-release courses. The format of the evolved schemes resulting in the present DH centrally-coordinated scheme is described in *Guidelines for vocational training for general dental practice.*[3]

Over the intervening years in the Trent region we have developed and refined our approach to dental vocational training, partly through the experience of providing courses and developing the dental vocational scheme and partly through the exploration of training in related fields such as medical general practice. Our experience of working with over 150 trainers and potential trainers in the Trent region and elsewhere suggests that many colleagues are interested in becoming trainers yet few have had the opportunity to think out what is involved in training and, in particular, in dental vocational training.

What is training?

Technically speaking, training is the provision of systematic structural opportunities to learn, usually with direct feedback.[4] Thus learning to fly an aircraft or to perform an operation on the temporomandibular joint involves training. More loosely, the term 'training' is used almost as if it were synonymous with education. However, training has the connotation of a strong vocational bias, whereas the term 'education' usually carries a connotation of broader cultural aims and the development of the whole person. Dental vocational training involves both training in the narrow sense and in the wider sense. Hence the methods used in dental vocational training can be seen as a broad spectrum of activities.

At one extreme is the lecture and at the other is independent study. In between is a rich variety of methods including small-group teaching, one-to-one tutorials, tightly structured practical tasks, observation tasks, more open-ended tasks, projects and independent practice. Trainer control is at its maximum in the lecture and trainee control is at its maximum in independent study. In both extremes, feedback to the trainee whilst learning is minimal.

The tasks of training may be concerned with the acquisition of knowledge, the deepening of understanding, the improvement of practical skills and the development of professional attitudes. Of these tasks, the development of professional attitudes is the most sensitive. Indeed, it is arguable whether a frontal attack upon attitudes is likely to be as effective as an indirect approach which focuses upon deepening understanding and improving practical skills.

As yet, there is little evidence on the effectiveness of dental vocational training in this country. There is evidence that medical vocational training has some effects. Cartwright and Anderson[5] show that 'doctors with a trainee year or a recognised vocational training scheme were more often regarded as easy to talk to by their patients and they were also more often felt to be good about visiting when asked to do so'. Freeman and Byrne[6] show improvements in clinical factual recall and problem-solving skills in patient management. Other studies of training indicate an improvement in practical skills and understanding, in medicine,[7] nursing,[8] social work,[9] management[10] and in teaching.[11] There is also evidence that small group discussions are an effective method of attitude exploration and change.[12] Finally there is evidence that trainers who themselves have been trained with feedback are better trainers than those who have not.[13]

However, one should be cautious at present about the overall impact of dental vocational schemes upon dental practice. Much will depend upon the nature of the schemes and the quality of training that the trainers and trainees receive. Hence, the importance of being clear about the nature of training and the tasks of trainers and trainees.

The purpose of dental vocational training

Dental vocational training involves a plethora of tasks which are all concerned with developing an independent competent professional who has a commitment to continue to learn. As such, its primary function is to provide in-depth professional guidance and its secondary function is to act as a bridge for the trainee between undergraduate and continuing postgraduate dental education. The primary function dictates that the bulk of training opportunities must be within the practice. It follows that the trainers are the key figures in the training process and that training the trainers is an essential step.

The focus upon training in practice leads to the importance of accurate feedback and assessment, of designing and providing learning tasks for the individual trainee. The dental vocational training scheme becomes trainee-centred and the trainer extends his or her professional role and expertise.

In providing training for trainers, one is also indirectly providing postgraduate dental education related to the needs of general dental practitioners. Again, research in dental vocational training is scant, but it is likely that through training others, one deepens one's understanding and develops one's expertise. Certainly, this view has long been known to educators through the maxim of Comenius[14] 'we learn as we teach'.

It would be wrong to leave this brief discussion of the nature of training without pointing out that all dental vocational training schemes have a conceptual basis, even if that conceptual basis is not fully recognised or understood by its designers and users. For example, consider a hypothetical vocational training scheme which focuses almost exclusively upon the 30 days of teaching received by trainees outside of the practice. Such an approach implies, perhaps unwittingly, that the priority of training is extending knowledge, not developing professional skills. This form of course may be assuming that training is merely telling and that accurate feedback and assessment in the dental practice setting are not important. Stressing the day-release course may undervalue the professional expertise of the trainers and it may appear to some members of the profession that dental vocational training is but a faint-hearted attempt at extending the undergraduate course. In contrast, when the dental vocational training scheme focuses upon learning in practice, then practice in learning becomes a priority and the scheme becomes both trainee-centred and training centred.

Some misconceptions of training

Discussions with dental colleagues on training courses often reveal anxieties and misunderstandings of the nature of training. Three recurring issues are training and personality, the nature of skills, and whether only dentists should be involved in the training of other dentists because only dentists understand the problems.

Training and personality

Two opposing views of training are that training will produce homogenised dentists and training cannot work because human beings are so different. The first view underestimates the resilience of human beings, the second underestimates their flexibility and responsiveness. Without a capacity for training, it is unlikely that there would be civilisation, let alone the dental profession. This is not to say that all forms of training are effective nor is it to claim that training can make an indifferent dentist into a brilliant one. It might, however, make one a better dentist.

Skills

Skills are sometimes taken to mean mere mechanical applications of practical procedures. Such a view debases the term, and the nature of skills and much of dentistry. Skills, by definition, are goal-directed sequences of behaviour. Activities are practised, learnt and become subroutines of actions. Performance, or stream of action, is controlled by feedback which may or may not be continuous. This input influences the selection of subroutines and matches the actions taken against the criteria of achievement or of goal approach. When engaged in skill performance, one's behaviour is directed and adaptive. Skills are disrupted when the feedback requires one to take actions that are unanticipated or unknown. If this occurs one can freeze, over-correct, ignore the feedback signals or respond appropriately. Feedback is a crucial element of skills learning. Inadequate feedback may well reinforce bad habits. Put another way, experience *per se* is no guarantee of effectiveness. Socrates is reputed to have said 'Experience teaches our best flautists. Unfortunately, it also teaches our worst'. But feedback, whilst crucial, is not enough. The feedback may reveal a gap between present performance and the goal. It is the trainer's task through guidance and support to close that gap.

This rather technical description of skills may seem remote

A graphical representation of how a dental consultation can be analysed into skill criteria. The scene: a patient enters the dental surgery complaining of pain and holding his hand over his right cheek.

Characteristic	Excerpts from consultation	Chairside example
Goal directed	D—What is the problem?	The diagnosis.
Learned subroutine	D—What causes the pain?	Questions and examination.
Routine feedback	P—Hot and cold.	Answers fit common diagnosis.
Disruptive feedback	P—Also when I touch my lip.	Answers suggest unusual feature. Further exploration necessary.
Possible errors	D—Sounds like trigeminal neuralgia.	Over cautious referral to hospital.
Correct action	D—Let me look and see if there is a dental cause. Discovers early perio abscess on canine.	Explore systematically through further questioning and inspection. Then decide.

from dental practice. Yet it applies to both operative and chairside skills. Often chairside and consultation skills are thought to be beyond training. 'One either has it or doesn't have it' is a frequently expressed opinion. Such a view is not wholly borne out by the evidence. As stated earlier, most people can be helped to improve their communication skills as well as their practical skills.

Contributions by other professionals and practitioners

The question is sometimes raised whether other professions can contribute to dental vocational training. The answer is a qualified yes. For example, psychologists have long been interested in practical and communication skills. Provided they take steps to understand the context and constraints in which dentists work, they can make a contribution—just as physiologists, microbiologists, accountants and colleagues from dental schools do.

The tasks of trainees

Put rather bluntly, the tasks of trainees are all concerned with improvement. This may involve the acquisition of additional knowledge, the refinement and extension of existing practical skills, the development and maintenance of relationships with patients and their relatives, and staff, colleagues and others. It may involve learning how to administer and manage equipment, materials and the resources of the practice and the associated tasks of record keeping and dealing with finance. These tasks may be classified along the lines advocated by cognitive psychologists[15] as:
● Accretion—knowledge acquisition, such as learning how and when to use a newly developed dental material.
● Tuning—refreshment of skills and their relevant knowledge base, such as review of the management of prosthetic problems in general practice.
● Restructuring—extensions and alternatives of previously acquired conceptions or misconceptions, such as 'one can train a dentist to operate but not to communicate'.

Restructuring is perhaps the most challenging because it requires a fundamental shift in thinking, perception and attitudes. Each type of task requires different training strategies. Accretion may be tackled by lectures and independent studies. Tuning requires guidance, feedback and assessment. Restructuring requires practical tasks with feedback, explanations and discussions, and may be helped by carrying out intellectually challenging tasks such as small research projects. All of the tasks require the trainees to assess, reflect upon and analyse their own learning. In so doing they are developing skills and strategies which will help them long after the training course has been completed.

Although we speak of trainees' tasks, in reality each trainee will have particular strengths and weaknesses which arise partly out of the trainee's own capacities and interests and partly out of his or her experience of undergraduate dental education. For whilst all dental schools necessarily follow GDC recommendations,[16] they do not teach identical courses. It follows that the training provided must be appropriate for the particular trainee. Clearly dental vocational training cannot be (nor should it be) a repeat of the undergraduate course, nor is the trainer-trainee relationship a disguised lecturer-student relationship. Rather the training seeks to build an individual strength and remedy individual weaknesses within the context of a partnership between a senior professional and a junior professional.

The tasks of trainers

The implications of the trainees' tasks for the tasks of trainers are clear. As well as being a highly competent professional, a trainer should have the capacity to develop a sound trainer-trainee relationship, to diagnose and assess all aspects of the trainee's work, to design and provide learning tasks, to give guidance and support and to provide accurate and meaningful feedback to the trainee. This view of the task of trainers is derived from the nature of training and the tasks of trainees. It is also largely in accord with the perceptions of trainees.[17]

At the risk of over-simplification there are four levels of tasks for trainer:
● Knowing how to use various procedures and strategies.
● Knowing why these various procedures and strategies are appropriate.
● Knowing how to train others in these procedures and strategies.
● Knowing why the approaches to training that he or she is using are appropriate for that task and trainee.

Clearly, the third and fourth level of tasks are challenging. They go beyond the notion of an experienced dentist acting as a good model to be imitated by the trainees. Rather than the trainee modelling the master trainer, the trainer has to help the trainee to master good models of practice. The tasks may require a trainer to consider and perhaps restructure his or her conception of training, to deepen understanding of the nature of assessment and to grasp that assessing another professional's work and approach is both legitimate and necessary as a trainer.

In carrying out these tasks, self-assessment, reflection and analysis by the trainer of his or her methods of practice are of importance. For these strategies provide the basis for developing the training of the trainees.

The challenge of assessment

A particular and frequently unwelcome challenge for the trainer is assessment and evaluation. Misconceptions, attitudes and self-analysis are closely bound up in the rationale of assessment in dental vocational training. Assessment is often misconceived as merely certifying a licence to practise. This form of assessment is known as certifying assessment.[18] Trainees and many trainers may not know the label but they have all experienced this assessment in 'finals', where it is used to protect society from the incompetent.

The other type of assessment is formative assessment.[19] As its name suggests, it is concerned with helping someone to improve by providing accurate, meaningful feedback of their progress. It measures the progress or gains made by the trainee from the moment he or she enters vocational training to the time he or she leaves it. The purpose is to inform the trainees of their level of performance, to provide an incentive to learning, to provide feedback on various parts in the course to show how a trainee's competencies are changing and to ascertain the quality of training itself.

Last but not least, formative assessment may be used to develop the trainee's own skills of self-assessment. Even if assessment is fully understood there may be a reluctance to use it since much of one's earlier training stresses that one should not criticise or even monitor the work of a fellow general dental practitioner. This attitude often hides under the banner 'professional good manners'. Hospital dentistry is different due to the long-established hierarchical training structure.

The matter of assessment is complicated further because there is a choice of approach to operational procedures and chairside skills. Some of these are correct, some incorrect and some may be correct but different from the habitual practice of the trainer. So the trainer may have to reflect upon his or her own practices to distinguish carefully between correct, correct but different, and incorrect approaches. Value questions are as much a part of dentistry as of any other profession. One of the tasks of the course organiser is to provide the opportunity for and encourage trainers to reflect on their own procedures and assumptions. The initial 2-day course for trainers provides this opportunity. Follow up sessions on topics such as using video in dental vocational training,[20] the consultational tutorial and the use of questions in dental vocational training all serve to help the trainer examine his own approach. These courses have stimulated the development of informal self-help groups of trainers who meet to propose and discuss various topics and approaches to training.

The 2-day course for trainers

The design of this 2-day residential course was based partly upon the previous sections of this chapter and partly upon the practical guidelines for course design in Brown and Atkins (1988) and in Miller (1987).[21] In essence these guidelines are: consider the goals, provide learning activities for the participants with tutor inputs and brief lectures, timing of parts of the course, the structure and assessment of the course.

As an aside, it is worth noting that studies of how people actually plan courses indicate that they zig zag from one problem to another rather than design in a linear fashion which begins with objectives and ends with outcomes.[22]

The course for trainers has as its primary goal the exploration by trainers of the nature and purposes of dental vocational training. A secondary goal was to enable new trainers to work together and develop a network of support.

The primary goal necessarily involved:
● Deepening understanding of the concepts of training.
● Analysis and appreciation of the tasks of trainees and trainers.
● The trainer-trainee relationship.
● The trainer-trainee tutorial.
● Designing and providing learning opportunities in the practice.
● Feedback and assessment.

Each of these components is worthy of at least a 2-day course in its own right. In a sense, the initial course is a seedbed for future work rather than a complete training package. A variety of methods of teaching and learning was used, including various practical tasks, discussions, video recordings of mini-tutorials given by the trainers, and brief lectures by the course tutors. The approach, as well as the content, was designed to help the trainers work with trainees.

The course began with an open forum in which the trainers introduced themselves and described their experience and interests in dentistry and training. This informal session was followed by a brief introduction to the nature of training and the first task of the course: to analyse the tasks of trainees. After comparing and classifying the material generated, the trainers were invited to analyse the tasks of trainers and in particular what a trainer might do in the early part of the training year.

At this stage, parts of the training video *Which trainer?*[23] are used to stimulate discussion. One of the trainers shown in the video is a highly competent dental operator but poor at personal relationships, one is probably capable but uninterested and the third is warm, friendly and sociable, but not very competent. The question addressed is which is the best trainer from the standpoint of a trainee and a course organiser. Discussion often centres on whether it is easier to develop a trainer's communications skills or his or her operating skills.

At the core of training is the trainer-trainee relationship and at its heart is the trainer-trainee tutorial. Participants are given a brief introduction to the types of tutorials and various approaches to preparation, to questioning, listening, responding, explaining and providing feedback. Common weaknesses in tutorials are identified. This introduction leads in to the task of identifying the characteristics of a 'good' training tutorial and a poor one. The criteria that they have generated are then implemented in the 'dental explanation game' which simulates the brief off-the-cuff tutorials that might occur over coffee or during a practical session. This not only gives the trainers experience in 'rapid response' tutorials, but also introduces them to the task of assessing the performance of a fellow professional and to the importance of having agreed criteria on which to base assessments.

The next session is an introduction to the formal tutorial. Participants working in small groups are invited to provide a brief, previously prepared presentation to their colleagues. These are video recorded. Each person takes the role of trainer, timekeeper, camera operator and trainee in turn. After the video recordings have been made, there is a lecture on giving explanations and tutorials. This input provides the basis for the analysis of the video presentations in which the presenter has to first assess his or her own performance and second, discuss the comments of other members of the group.

The video sessions provide a springboard for the introduction and discussion of self-assessment and assessment. These sessions also exemplify the collection and use of evidence to make informed judgements. Three types of evidence are identified: evidence from direct observations of the trainee at work, evidence from reported observations by patients, staff and colleagues and evidence deduced from later examination of the trainee's work or patients' reactions. In this session the trainers are also introduced to the joint assessment procedures given in the log book for trainees.

The log book leads into the final session of the course which is concerned with how individuals differ in their approaches to learning. The participants are set a learning task and assessed on it. They are asked to discuss how they tackled the task and how they felt whilst tackling it. This discussion leads to a formal presentation of the research on how adults, particularly students, learn and to the various styles of learning which have been identified.[24] The trainers are then asked to identify their own styles of learning and discuss the conditions under which they learn well and badly. The implications for their provision of training for trainees are drawn out. The key message is 'know your trainee's tasks, know your trainee's approach to learning'.

The course ends with a brief session on course design and the evaluation of the 2-day course, using rating schedules provided, and an invitation for open-ended comments.

Discerning readers may have noted that the course does not follow a logical structure based on the subject of training nor is it based on lectures. A logical approach which begins with definitions and methods is not necessarily the best psychological approach to a subject as complex and sensitive as training. Nor are formal lectures the best method of deepening understanding and developing practical expertise[24]. Information transmitted is not necessarily information received and understood. Understanding the practical skills of training requires opportunities for exploration, practice with feedback and structured discussions.

The 2-day training course received favourable evaluations which will be the subject of chapter 29. Whilst these comments and ratings, as such, do not prove that the course for trainers was effective, they do provide strong evidence of a worthwhile course, trainer satisfaction and the prospect of a group of motivated trainers. The essential steps have been taken.

References

1 Fordyce G L, Brookman D J, Kilty J M. Training for general dental practice: a pilot experiment. *Br Dent J* 1977; **143**: 156-159.
2 Conference of Postgraduate Dental Deans. Vocational training: the nuts and bolts. British Postgraduate Medical Federation, 1982.
3 Brookman D J, Horrocks J K, Lowndes P R, Pendlebury M E, Robb I M. *Guidelines for vocational training for general dental practice.* UK Conference of Advisers and Associate Advisers in General Dental Practice. British Postgraduate Medical Federation, 1987.
4 Holdin E. *Principles of training.* London: Pergamon, 1961. Stammers R,

Patrick J. *The psychology of training.* London: Methuen, 1977.
5 Cartwright A, Anderson R. *General practice revisited.* ppl45. London: Tavistock Publications, 1981.
6 Freeman J A, Byrne P. Clinical factual recall and patient management skill in general practice. *Medical Education*; **11**: 39-47.
7 Maguire P, Fairbairn S, Fletcher C. Consultation skills of young doctors. Benefits of feedback in interviewing. *Br Med J* 1986; **292**: 1573-1576.
8 Davis B. Social skills and nursing *In* Argyle M (ed). *Social skills and health.* London: Methuen, 1981. Faulkner A, Maguire P. Teaching ward resources to monitor learner patients. *Clinical Oncology* 1984; **10**: 383-389.
9 Hudson B. Social work training. In Argyle M (ed). *Social skills and health.* London: Methuen, 1981 .
10 Philips K, Fraser T. *The management of interpersonal skills training.* Hampshire: Gower Press, 1982.
11 Brown G, Shaw M. Social skilled training in education. *In* Hollins C, Trower P (eds). *Handbook of social skills training.* Oxford: Pergamon Press, 1986.
12 Abercrombie M L J. *The anatomy of judgement.* London: Hutchinson, 1960. Jaques D. *Small group teaching.* London: Croom Helm, 1984.
13 Naji S A, Fairbairn S A, Maguire P, Goldsberg D P, Faragher E B. Training clinical teachers in psychiatry to teach interviewing skills to medical students. *Medical Education* 1986; **20**: 140-147.
14 Comenius J A. *Didactica Magna.* Amsterdam: Laurentii de Greet, 1697.
15 Norman D. Cognitive engineering in education. *In* Tumo D J, Reis S (eds). *Problem solving and education.* New Jersey: Lawrence Erlbaum, 1980.
16 Recommendations concerning the dental curriculum. London: General Dental Council, May 1985.
17 See chapter 29.
18 Downie N M. *Fundamentals of measurement: Techniques and practices.* New York: Oxford University Press, 1967.
19 Brown G, Atkins M. *Testing and learning.* London: Longmans, 1985. Harris D, Bell C. *Evaluating and assessing for learning.* London: Kogan Page, 1986.
20 Pendlebury M E, Brown G. How to use video in dental vocational training. Nottingham: Queen's Medical Centre, 1987.
21 Miller A. *Course design for university lecturers.* London: Kogan Page, 1987.
22 Wittrock M. Teachers cognitive processes. *In* Wittrock M (ed). *Handbook of research in teaching.* New York: Macmillan, 1985.
23 Pendlebury M E, Brown G. *Which trainer. . . ?* Nottingham: Queen's Medical Centre, 1987.
24 Brown G, Atkins M. *Effective teaching in higher education.* London: Methuen, 1988.

29

Trainees' Perceptions of the Tasks of the Trainer

M. E. Pendlebury

In courses designed for training trainers in the Trent region, potential trainers are asked to define the tasks of new graduates entering vocational training for dental practice. They then explore and attempt to define their role as trainers, matching their perceptions of the needs of the trainees,[1] and also following the design path of the initial vocational training courses.[2] It was decided to explore the trainees' perceptions and valued characteristics of vocational training and trainers. Such information could then be used in the design of subsequent courses for trainers and trainees and also as a contribution to the course for trainers.

The beginning of the 1988 scheme was considered opportune, since Trent now has two schemes, which are both integrated with the community service. This sample contains 15·8% (133 total)[3] of the general practice vocational trainees in England and Wales starting on the new DH schemes in January 1988. The study was carried out during the first 6 weeks of their vocational training.

The trainees

Twenty-seven graduates entered dental vocational training in Trent region and attended day-release courses based either in Sheffield, starting in the last week of January, or in Nottingham, starting in the first week in February (Table 1). The trainees had started work in their practices before these courses began. The majority had qualified in December, but three had qualified earlier in 1987 and one had qualified in December 1986 but had only recently entered general dental practice.

The trainees came from eight dental schools, with the Sheffield course demonstrating the well-known phenomenon of graduates clustering around their school of origin.[4] The Nottingham course, not being associated with a dental school locally, shows a greater mix of origins (Table II).

On the first day of the course, the Nottingham group of trainees were asked to spend a few minutes considering the tasks of their trainers during vocational training, and they then formed into two groups to discuss and complete a short list of factors thought relevant. This produced two similar lists (Table III) and a master was compiled, based on all the elements occurring in the trainees' lists, with the suggestions not arranged in any priority order.

A questionnaire was developed from this list, which sought two separate opinions. First, each trainee had to suggest a priority order and secondly, they had to indicate the level of agreement with each short statement on a four-point scale. The four points were 'strongly agree', 'agree', 'disagree' or 'strongly disagree' with the statements.

Twenty-five of the trainees spent 2 days at British Dental Association headquarters, the General Dental Council and the then Dental Estimates Board. One community trainee did not attend and one general practice trainee had left the scheme. The questionnaire was issued on the first day of this visit and collected on completion. All the trainees returned the completed questionnaires.

Priorities and strength of agreement

The overall priority of the tasks was calculated by summating

each trainees' rank order. This procedure provided a profile of the trainees' perceived tasks of the trainer. They are given in ascending order in Table IV. The strength of agreements with the statements are shown in Table V. Strong agreement and agreement were assigned scores of 1 and 2, respectively, and −1 and −2 for disagree and strongly disagree. This enabled a double check to be made upon the views of the trainees.

Figure 1 shows the priorities when scores on the four-point scale are summated algebraically. The overall order of

Table I Dental vocational trainees in the Trent region

Centre	General	Community	Total
Nottingham	13	2	15
Sheffield	9	3	12
Total			27

Table II Trainees' dental schools

	Nottingham course	Sheffield course
Liverpool	3	1
Leeds	5	0
Sheffield	1	6
Newcastle	0	2
Manchester	2	2
Birmingham	1	1
Bristol	1	0
Dundee	2	0

Table III Tasks of the trainer

Sub-group 1	Sub-group 2
1 Guidance in management—patients, practice, staff, etc	1 Practice management
2 Teacher and tutor	2 Advice/help when needed
3 Support	3 Sympathetic
4 Problem solvers	4 Enlarge range of experience
5 Friend and colleague	5 Facilities
6 Be available/aware	6 Help the trainee to integrate socially
7 Be there when needed without being intimidating/interfering	

Table IV Trainees' expectations as expressed priorities

9th	Help the trainee to integrate socially
8th	Act as a problem solver
7th	Be a teacher and tutor
6th	Be a friend and colleague
5th	Should be sympathetic
4th	Plan to enlarge the trainee's experience
3rd	Should be approachable
2nd	Guidance in management of patients, practice, staff, etc
1st	Be there when needed without being intimidating/interfering

The Questionnaire

State Priority	Tasks of the trainer	Strongly agree	Agree	Disagree	Strongly disagree
	To give guidance in management of patients, practice, staff, etc				
	To be a teacher and tutor				
	To act as a problem solver				
	To be a friend and colleague				
	To be there when needed without being intimidating and/or interfering				
	Should be sympathetic				
	Plan to enlarge the range of the trainee's experience				
	Help the trainee to integrate socially				
	Should be approachable				

priorities is closely similar to the rank order (Spearmans rho = + 1). 'Friend and colleague' and 'Teacher and tutor' changed rank order as did 'Be approachable' and 'Give guidance'. Figure 2 shows the rank order of valued tasks when agreements only are considered. Again the priorities are consistent with the initial priorities (Spearmans rho = + 1). It may, therefore, be concluded that the survey accurately represents this sample's priorities of trainer tasks. The highest priorities were, in descending order,
● be there without being intimidating or interfering;
● guidance in management of patients, practice, staff, etc;
● be approachable;
● plan to enlarge trainee's range of experience .

The very low priority of 'Help the trainee integrate socially' could be attributable to the relatively high proportion of Sheffield graduates who presumably would have already developed a strong social structure. However, trainees who are new to an area would probably welcome some help towards social integration. Interestingly, it was the Nottingham group who suggested this factor as one of importance.

Clearly, the trainees in the sample did not value highly the task of a trainer as a 'teacher and tutor'. In discussion with trainees, following analysis of the questionnaire, one trainee stated 'it would be unwise to call them teacher and tutor as this could give them ideas above their station'.

It might also be that the trainees' perception of 'teacher and tutor' was based on an implicit model of the lecturer-student relationship. Comments from other trainees suggested that, in the minds of trainees, teaching consists of force feeding

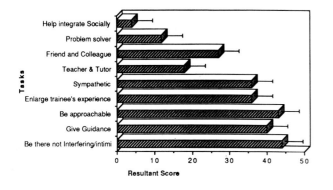

Fig. 1   Agreements with disagreements deducted.

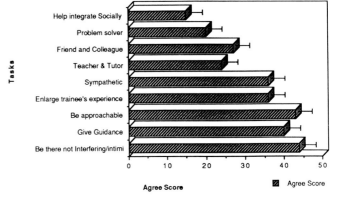

Fig. 2   Preference from agreements.

information, whether that information was needed or not. The trainees' concept of teaching and tutoring and perhaps of learning requires some further exploration.[5]

It would appear that rather than overt direction from a trainer, the trainee is expecting the trainer to act as a 'life-belt' which can be called on when needed.

A task of the trainer that was not highlighted by this group of new trainees was assessment. Yet assessment is essential if a trainer is to be aware, to provide guidance and to enlarge experience. It will be interesting to see whether this group's perceptions of the tasks of the trainer change through their experience of vocational training. Equally interesting is whether the training undertaken by trainers will assist them in their tasks, identified by this group of trainees.

Departure from the lecturer-student model

The study indicates that a sample of vocational trainees entering vocational training have a similar view on the tasks of the trainer. They consider the four major tasks are for the trainer to be aware but not intimidating or interfering, to give guidance, to be approachable and to plan to enlarge the trainee's experience. Lowest priorities were assigned to being a teacher and tutor, acting as a problem solver and helping to integrate the trainee socially. The trainees' concepts of the nature of training did not appear to include the notion of assessment. These findings suggest that trainers might profitably explore with trainees their views on the tasks of the trainer and that trainers should develop approaches to training that are different from the implicit lecturer-student model.

References

1 See chapter 28.
2 Fordyce G L. Vocational training for general practice. The pilot scheme. *Dental Update* 1985; **12:** 379-380, 382-385.
3 Committee on Vocational Training, 33 Millman Street, London. Figures as at March 29, 1988.
4 Thexton A, McGarrick J D. The geographical distribution of recently qualified dental graduates (1975-80) in England, Wales and Scotland. *Br Dent J* 1983; **154:** 71-76 .
5 Pendlebury M E, Brown G A. A course for trainers in Trent. DVTS: Queen's Medical Centre, 1988.

30

Preparation for Working Together

J. K. Horrocks

The trainer will have a greater influence on his/her trainee during the training period than any other single person or event. In addition to being the trainee's role model as a practitioner, the trainer will be colleague, guide, adviser, supervisor, mentor . . . and maybe even friend! A good working relationship between them is essential if this influence is to benefit the trainee's development and maturation. This is a personal view of some of the issues involved in establishing from the outset an effective working relationship between trainers and trainees.

Newly-qualified graduates and dentists entering general dental practice for the first time want to become competent as soon as possible in functioning independently in this, their chosen specialty. The main purpose of vocational training is to help them to achieve this goal more surely and more quickly than they could on their own, but it is our experience that new graduates are often unclear as to what to expect of vocational training. Some have confessed to believing it to be an easy ride! We have also found that experienced practitioners do not necessarily become effective tutors and supervisors merely by the act of approval as trainers and a brief induction.

Perhaps the main task for those responsible for vocational training schemes is to help trainers and trainees to define their goals and to develop the skills necessary to attain them. However, it is necessary to consider where they start from, before it is possible to create an atmosphere in which both may learn.

> 'If I don't know I don't know
> I think I know
> If I don't know I know
> I think I don't know. '
>
> (*Knots*, R. D. Laing)

The trainees

The new graduate's entry to practice is one of her or his major upheavals in life, from being top of the pile, at last, to being bottom of the heap, again. The longed-for day of graduation is followed by the grey dawn of a new day in a very different world: how to survive, how to cope, and then, of course, how to succeed.

The trainees have emerged from a long period of formal education and may not yet have realised that they are entering a period of a different kind of educational experience, in which the familiar hierarchy of knowledge, skills and attitudes is reversed. This is a period during which they need to become active in their own learning rather than being taught, and become convinced of the value of continuing education. As they transfer the knowledge and skills they already have to general practice, they will need to welcome criticism in order to become self-critical. If they are to acquire and develop the attributes necessary to succeed in practice, the value of the experience of the 'ordinary' dentist must be emphasised.

The trainers

The trainers have lived through all this. They not only survived, but succeeded. Most of them learned by trial and error, probably with some pain and tribulation. It should not be surprising if they have largely internalised the attitudes, skills and knowledge they have acquired over the years, nor that they may find difficulty in becoming convinced that those attributes can be defined and acquired more quickly and surely by an organised educational process.

Now the newly-approved trainer is experiencing another great change, from successful, experienced practitioner to a new, unfamiliar role. For it is one thing to do, quite another to help somebody else learn how! For them, experience was perhaps their only tutor, and they may not have formally developed teaching skills.

Facing up to change

The common response sequence to a change or transition is first disbelief, then numbness, followed by depression and finally acceptance of change. New trainers and trainees both experience such a cycle of change in their self-esteem. I feel that this should be addressed from the outset and that it is important to bring them to acknowledge that it is a normal sequence. I believe they can be helped to move quickly from feelings of 'this can't be what I wanted' through 'if I ignore it, perhaps it won't happen' and 'I'm not coping at all' to 'I'm enjoying this and doing well' . If the negative feelings of transition are addressed successfully, it will be possible to develop a sound working relationship.

I believe that an effective working relationship is one based on value, respect, trust and honesty. Thus, trainers and trainees need to value each other, first as people, then as professionals and dentists. They need to bring to the relationship mutual respect for what the other has to offer and to acknowledge that each has something to learn from the other. They should be honest and open, able to trust that what they say and do together is in confidence. Finally, they need jointly to define their goals and ensure that both believe these to be relevant, valuable and attainable. Trainers will probably realise this for they are experienced in establishing just such relationships with colleagues, staff and patients. It may be a new concept to trainees, who might not initially see its relevance to their professional and clinical life.

It seemed logical to start the training year by providing an opportunity for the trainers and trainees to understand one another's expectations and to explore how they will work together in practice and in tutorials. At this stage, two of us joined together to plan the programme of introduction to our vocational training schemes. We decided to offer a residential course in relaxed comfortable surroundings. Living together would give them a shared experience to start and sustain their year. Away from the practice they would be able to see and value each other, first as people. Bleary-eyed contact across the breakfast table is a useful extension to the common 'trial by wine'.

On course
They would work together first on non-dental exercises designed to help them examine their personal values in a supportive atmosphere in which all could participate freely.

They would have the opportunity to talk about their expectations—of each other, the course, the year. They would see that they all had something to learn and something to bring to the year. They would discuss ways of analysing the trainees' needs and how to meet them by daily support and supervision. They would practise tutorial techniques and be helped to bring out the knowledge each possessed so that both could gain from the experience. Finally, they would be helped to see the value of periodic assessment and final evaluation, to check regularly that they were on target and, if necessary, to make adjustments in order to achieve their goals.

31

Beginning To Work Together

J. K. Horrocks and W. R. Allen

Ten people met for the first time as trainees after lunch on Thursday, February 4, 1988 at an hotel near Ipswich. They were to live and work together until teatime on Saturday. Twelve trainers joined them from late afternoon on Friday, staying on after the trainees left, until lunchtime on the Sunday. The occasion was the start of the 1988 vocational training schemes for East Anglia and North East Thames regions.

Experienced practitioners are expert in the provision of dental care, in the management of personnel and resources, business finance and administration. Their task as trainers involves assisting trainees to acquire and develop this body of knowledge and skills which they have largely internalised. This transfer of experience should not be left to chance.

It is convenient to regard the year as a journey they are to travel together. Wise travellers plan their journey with care, in order to arrive at their chosen destination. They decide where they want to go, and why, and they consider their departure points—'If I wanted to go there, I wouldn't have started from here,' goes the Suffolk saying—and work out the route and means of travel. Along the way they will use landmarks to check that they are on the right road, and on schedule; finally, they will confirm that they have arrived at the place they intended to be.

This introduction course was the start of the journey and we, the course organisers, were the guides engaged to help self-directing adults to make responsible and realistic assessments of their own progress towards their goals. The first stage of this journey aimed to identify their needs, define their goals and agree the route. Our secondary aims for the weekend were to help the trainees to develop a clear understanding of vocational training, and to assist them and the trainers in the development of a group ethos.

The trainees had recently graduated from some of the London schools, as well as from Edinburgh, Newcastle, and Sheffield, and so were able to bring to the year a rich diversity of academic knowledge and opinion. The trainers came from practices in London, Metropolitan and North Essex, Cambridgeshire, Norfolk, and Suffolk. The 'country' practices were situated in such lovely towns as Braintree, Halesworth and Saxmundham, as well as busy, thriving Ipswich, Peterborough and Yarmouth.

Everyone knows where London is, but not everyone seems to know that Essex and East Anglia is that bit of England sticking out to the right, between the Thames and the Wash. It is a rapidly-growing 'boom' area, with low unemployment and high income potential for the expanding population. Sadly, some of our new graduates are not aware of the attractions of the area and two of the trainers were unable to attract trainees to their practices.

After celebrating the success of 'finals', new graduates need some time to find jobs with dentists who may only recently have been approved as trainers. Christmas festivities intervene and we have found that it is the beginning of February before most trainees are in post and we can start their training year.

The Committee on Vocational Training provides that most of the taught course takes place during the normal working week because of the method of funding, and we believed that our timing of the introduction would satisfy this requirement, yet still allow a break before returning to practice the following week. We chose the Ipswich Moat House Hotel as our venue, because it offered good accommodation, food and facilities at reasonable cost, and was mutually inconvenient to most of us, being situated on the main London to Yarmouth road.

'Tell me and I forget, teach me and I remember, involve me and I learn' sums up our philosophy of involving everyone in their own learning from the start; educational methods we use to facilitate this include discussions, short lectures, role-play, and experiential group tasks.

Course diary
When the trainees arrived after lunch on Thursday, we set the scene for an informal, participative working group. The names we wished to be known by were clearly displayed on tabs as we sat in a circle and introduced ourselves. Each person shared a self-description with a neighbour, who then introduced them to the group. Thus, everyone named and established eye-contact with everyone else and, as the group relaxed and joked, we started to develop listening and feedback skills.

Large posters clearly displaying the aims and objectives of the weekend and of each session were explained and expanded upon. We wanted to establish at an early stage the habit of defining and stating our purpose as this would facilitate future assessment. Clearly spelling out what we were doing not only seemed to be appreciated, but also reduced the myth of the 'hidden agenda'.

Volunteers were given the tasks of discovering the details of the domestic and administrative arrangements for the weekend and ensuring that the whole group got the information. This involved the members in active participation from the beginning, and established that members would not be spoon-fed during the year.

After a break for tea and some informal chat we moved on to identify the year's tasks for trainees and trainers, and to give an opportunity to express and discuss their expectations. An overview of the year was given, the 'taught course' defined and linked with practice experience. By relating this to their expectations and making it clear that they would be given increasing responsibility for the design of the year to meet their own needs, we continued the theme of active participation in the management of their own affairs, as is appropriate to self-directing adults.

After dinner that evening, Bernie Mayston, a trainee from

a previous year, described his experiences during his training year. A question and answer session followed, with each member being given 2 minutes to elicit as much information as possible on his nominated topic, a useful introduction to information-gathering in the surgery. So the day ended with the new trainees having been involved from the start.

Dental students, in our experience, have little opportunity to learn how to look after and maintain the equipment, instruments, and materials with which they work. Dental practitioners rely for their livelihoods on these items being in good order, and the first session on Friday morning was devoted to these practical and essential matters, led by Ken Williams, a local practitioner who was co-author of a *BDJ* booklet on the subject.

Similarly, as undergraduates the trainees will probably not formally have considered the components of interpersonal relationships in a working environment. The rest of the morning focused on relationships between people within the practice. This set the scene for the afternoon's work on devising strategies and tactics for forming and fostering the relationships necessary for the trainee's surgery unit to function within their training practices. This session was led by Barry Turner, a GDP and dental tutor, and went on to consider what a training practice might reasonably expect from a trainee in not only a professional and clinical sense, but also a personal one.

Non-dental exercises

Meanwhile the trainers had arrived and the course organisers repeated the welcome and introduction exercise, bringing them up to date with the course thus far. Trainers and trainees met as a group for the first time over dinner, and went on to work together on a non-dental exercise, designed to encourage them to consider their individual values of people and to develop negotiation skills.

Saturday began with another relevant non-dental exercise, which focused on the need for clear communication in the professional setting. This set the scene for a review of the trainees' expectations and their understanding of what would be expected of them by their practices. The group was divided into 'mixed pairs', of trainer and trainee, and moved on to review the trainees' expressed objectives for their first week in practice, discuss arrangements for weekly tutorial sessions in practice, and consider the mechanics of observing each other at work. The fact that each member was not in the same discussion pair as his own trainee or trainer allowed them to consider each other's thoughts in a non-threatening manner.

During the training year, trainers and trainees join together each week for a minimum of one hour's tutorial and supervision session. Exchanging stories of recent occurrences in the surgery can be mutually useful but we believe that the educational value of these sessions is greatly enhanced by agreeing discussion topics in advance and by preparing thoughts or work, in order to ensure that both parties bring something to the occasion.

To give members the opportunity to experience how it feels to be a learner or a teacher—how much easier it is to tell others how to do it!—the group was divided into subgroups of three who spent the afternoon enacting tutorial sessions, taking it in turns to role-play the part of trainer, trainee, and observer. Once again, no one was in the same trio as his own

trainer or trainee; from the feedback, we gathered that the session was well-received. The trainees departed at the end of the afternoon; by now we were quite tired and so spent a relaxed and informal evening with the trainers.

On Sunday, it was back to basics! Trent region produced a video—*Which trainer?*—caricaturing three very different types of trainer. We showed this to stimulate discussion on how best to introduce trainees to their practices before we moved on to consider the attributes of a 'good' dental practitioner. Then came a review of the aims of the training year, followed by thinking about the skills that experienced practitioners already possess that are transferable to their role as trainers and how they could develop the additional skills they needed. Finally, we summarised and discussed the experiences of the weekend, before enjoying lunch together and leaving for home.

Each session and day was debriefed by the leaders and the programme modified as appropriate. We had some difficulty due to the fact that some of the trainers had previous experience; some had been to preselection courses, some had not; some attended only part of this introductory course. Devising a programme that will adequately accommodate people at differing stages is almost impossible. Some of the non-dental exercises were not perceived as relevant by some of the course members, although we hoped that in retrospect they would understand why we had attempted to consider some aspects in a wider sense.

The course was assessed by the participants by asking them to score each session under the headings of 'interest', 'usefulness', 'organisation'. On the whole the scores were 'good' or 'very good'. Participants were also asked to add some general comments to help us in the future. Some of the reactions from the trainees: 'generally well-run and enjoyable'; 'good idea for trainees and trainers to work together'. And from the trainers: 'I think you get a commitment to the group at a residential course'; 'I feel I know exactly what is expected and I am much less daunted'; 'more groupwork, less theory' (wanted? or what?); 'role-play in groups of three is an ideal way to get across matters of experience'.

For our part, we felt that the course had been intensive and hard work. We felt we had gone a long way to achieving our aims and objectives and, although some modification will be necessary in the future, we feel it will be worthwhile to do again. Our original purpose in getting trainers and trainees together had been justified in the positive feelings that all expressed about this method of induction and we feel that it is a useful evolution of the vocational training scheme that we will wish to develop further.

Conclusion: 1990

Two years on, vocational training in the two regions has grown to the extent that a single joint introductory course is no longer practicable. Both regions have kept to the original philosophy of introduction to the training year. The style of presentation and the course content is continually modified and refined in response to feedback from the participants.

We are pleased that this feedback is overwhelmingly positive, confirming our belief in the value of a residential introduction which focuses on trainers and trainees as people bringing to the year attributes of equal worth.

32

Training the Trainers

P. R. Lowndes, R. J. K. Caddick and J. W. Frame

The continued success of the national vocational training scheme in dentistry depends largely on the commitment and expertise of the trainers in general dental practice. Training the future trainers is an important consideration. This chapter reviews the approach in the West Midlands.

A general dental practitioner is selected by his Regional Postgraduate Dental Education Committee to become a trainer because of his professional ability and integrity, high practice standards, commitment to provide a wide range of treatment, postgraduate study record and enthusiasm for the vocational training scheme. This appointment implies a responsibility to supervise the transition of a new graduate from the protected environment of a dental school to independent practice.

The relationship between the trainer and the trainee is more complex and interdependent than that between principal and associate. The trainer is responsible for the trainee's errors and omissions by virtue of the assistantship basis of the trainee's contract. In addition, the trainer has a formal teaching and monitoring role in relation to the trainee. The new trainer may have had little previous experience or instruction in teaching methods. Consequently it is important that the trainer is adequately prepared for this new role.

During the last 8 years in the West Midlands, trainer preparation has evolved to include:
- Participation in a residential trainer preparation course prior to the training year. (It includes medico-legal aspects, trainer's responsibilities, role-play and workshop sessions, and preparing for the practice seminar.)
- Meetings of trainers with the regional advisers prior to each term.
- Participation in the trainees' day-release courses.
- Attendance at a residential trainer/trainee conference.
- Advice on monitoring and assessing the trainee's progress.

The trainer preparation course

The structure and duration of such courses vary considerably from region to region. The present format in the West Midlands was established following feedback from trainers who had attended the earlier courses, with special attention being paid to timing. The course is run a few months before the start of the training year to allow adequate time for the trainers to assimilate all the information. The group meets in the early evening for dinner, followed by a working session. The trainers stay overnight at the conference centre and the course continues for the whole of the following day. This is felt to be the minimum time in which the relevant material can be adequately discussed. The overnight stay helps develop a group ethos.

The course is designed to illustrate the problems of being a trainer and adapting to the presence of a trainee in the practice. Trainers are encouraged to examine these problems and, as a group, generate their own solutions. Some didactic input is involved, mainly in explaining the techniques of one-to-one teaching and in examining the medico-legal implications. Active participation of the trainers in the course is regarded as important, with small-group educational methods being employed rather than a lecture-style approach.

Trainer's responsibilities

Groups fully discuss the medico-legal aspects of vocational training including the new trainer-trainee contract. The trainee is employed as an assistant in the practice and the trainer is responsible for the errors and omissions of the trainee as they relate to his or her contract with the Family Practitioner Committee. The trainee has direct responsibility to the patient and is therefore insured with a defence society and answerable to the General Dental Council on any disciplinary matters.

The central role and responsibilities of the general practice trainer in the new scheme are fully discussed by the regional advisers. The trainer should see himself as a role model for the trainee. In addition to providing the physical environment and staff support, the trainer should be eager to give clinical advice and assistance. There is also a more formal teaching role, involving the development of a structured weekly discussion programme. The trainer should ensure that the work load is appropriate, assess progress and undertake a clinical review.

The trainer and trainee assessment forms developed in the West Midlands are introduced at this stage as an aid to monitoring the progress of the trainee (see p.142). The completion of these forms at regular intervals is regarded as essential, both as a guide to progress and as an opportunity for the trainer and trainee to communicate by completing them jointly.

The remainder of the course takes the form of role-play sessions and advice is given on educational methods.

The role of role-play

In role-play a situation is demonstrated by providing group members with role cards indicating the personality and situation they are to portray and allowing them to develop the role spontaneously. This technique has been criticised as being too artificial and contrived, but it has been found to be an effective method of illustrating problems. It is noticeable that as members of a group repeat different role-plays they become less self-conscious and adapt more rapidly to scenarios. This method has been used to explore a number of problems in communication and assessment of standards.

Workshop sessions

In workshop sessions participants consider a problem in

Trainee Assessment Form

TRAINER   TRAINEE

....

TERM ....   DATE COMPLETED ....

Grades	1	2	3	4	5	Improve/Decline (+ / −)
Knowledge						
Existing						
Attitude to improvement						
Skills						
Diagnosis						
Treatment planning						
Conservation						
Prosthetics						
Crown and bridge work						
Oral surgery						
Periodontics						
Other						
Personal						
Reliability						
Attitude to patients						
Attitude to staff						

Key
1   Exceptionally good, marvellous, superlative (Clinical skill = excellent).
2   Good, above average, clever, industrious, sympathetic (Clinical skill = good).
3   Competent, adequate, dependable, pleasant individual (Clinical skill = average).
4   Lacks awareness, tends to irritate people, does not pull his or her weight, ...
    any of the above .... BUT would be worth sorting out (Clinical skills = poor).
5   Dreadful. A mistake in appointing as a trainee (Clinical skills = abysmal).

subsets, which ideally should confer in isolation. After an appropriate period, the subsets reconvene and report in rotation. The initial group gives a full report and the following groups relate only divergences. Finally, a general discussion ensures that the problem is thoroughly explored and the various options established. The exercise can be repeated with different problems and the membership of the subsets and their order of reporting varied. Once again, as the exercise is repeated, the response increases and individual involvement is enhanced.

Preparation for the practice seminar

The practice seminar is regarded as an important educational event in the vocational training programme. Advice on this has been received from Mr Bill Fleming, director of the Department of Medical Education at Birmingham University.

There are four aspects of preparation for the practice seminar: venue, time, personnel involved and the actual event. The seminar should be approached systematically. A non-threatening, interruption-free venue is required. The timing should be regular, so that it is a constant feature of the practice week and should be of one hour's duration. The main personnel are the trainer and trainee, but there is the option of involving others from inside and outside the practice. This should occur only occasionally to avoid dilution of the trainer/

trainee relationship. The event itself requires an agenda and an overall framework each term, although this should be flexible enough to include unexpected items derived from the practising week.

The trainee's clinical work is an important source of material for use in the seminars, and case notes, radiographs and study models are required. It might be useful to consider the concept of the critical incident, where the trainee presents an event that has gone well and an event that has not gone so well during the previous week. One should consider the setting, what happened, the outcome and if the result was effective or ineffective. Discussion of these factors may lead to congratulation or commiseration, with advice on avoiding and managing such problems in the future. The monitoring of critical incidents is a means of assessing the development of the trainee during the year.

Feedback after the course
A questionnaire is circulated to the participants to evaluate the course and to aid future developments. Although of short duration and thus highly concentrated, the general feeling has been that sufficient time was available to allow adequate coverage of the relevant subjects. Consideration is being given to running an extra evening course a few months after the trainees start in practice to discuss problems which might have arisen. Additional topics that have been suggested for inclusion in future courses are the financial aspects of having a trainee in the practice and the development of interviewing techniques for trainee selection.

The termly meeting with regional advisers
Prior to each term, trainers meet regional advisers at the postgraduate centres where the day-release courses are held. This provides an opportunity for a review of progress and exchange of views. The programme for the forthcoming term is distributed prior to the meeting and is explained in detail. The trainers are able to outline the practice seminars they have planned. These are normally related to the overall day-release programme. If the agenda is not too onerous, it is sometimes possible to arrange a speaker on a topic of general interest, usually related to the techniques of 'in-practice' teaching.

Participation in the trainees' day-release courses
Trainer participation in day-release courses takes two forms. First, trainers are welcome as group members at any session. It has been found that they act as secondary catalysts in the discussions and have a valuable effect in enlivening the sessions.

Secondly, many trainers have special skills and experience in the general practice field. Vocational training is an introduction to general dental practice and it is therefore appropriate that the majority of sessions are conducted by practitioners from this background. Trainers are encouraged to undertake this role, and suitable briefs and detailed support in audio-visual presentations are provided if required.

Trainer/trainee conference
It has been the custom in the West Midlands to arrange a conference for trainers and trainees each autumn, providing an opportunity to bring together all those involved in the scheme. The format is similar to the trainer preparation course. Members assemble for dinner, work an evening session and stay overnight, continuing the whole of the following day. It greatly enhances group ethos and again allows an exchange of views and discussion of problems.

Each conference has a coherent theme, often from the socio-economic aspects of general practice. Recent conferences have considered communication, discovering the patient, a study of patient attitudes to dental care, and ethics for the practitioner and the profession.

Future patterns of trainer preparation
Both locally and nationally, it has been observed that trainees tend to remain in their training practices at the end of the vocational training year, often staying for prolonged periods. This is in contrast to medical vocational training, where the post in practice is for one year, after which the doctor is replaced with a new trainee. In dentistry, the effect is to deny access to existing training practices until the trainee leaves or there is some other alteration in the practice personnel. The number of training practices required is therefore many more than the number of trainees in a given year and there is a continuing need to provide training for this large body of trainers.

Perhaps with the enhanced financial rewards now available to trainers, practices might emerge with a recurring commitment to take a new trainee each year on the medical model. This would allow the establishment of a core of experienced training practices, which might well influence future trainer preparation programmes.

The development of vocational training in dental practice is on-going and there will be changes and alteration in emphasis in the future. It is a challenging field, with much to offer and seek from trainers, trainees and course organisers. The rewards of the scheme's success will be higher standards of patient care and improvement in the quality of general dental practice in the United Kingdom.

33

Out of the Egg

C. Mills

That well-known dentist, General MacArthur, once said: 'There is no security on this earth, only opportunity'. The question I had to consider after graduation was whether to take the opportunity which presented itself to me, and to accept a place on the new national vocational training scheme.

I had long since realised that when I qualified a number of problems would confront me before I could take my first faltering steps into the world of the 'real' dentist. I was faced with the task of being expected to choose my first job, on which so much depends, without much of the knowledge and skill that is necessary to make a sound decision. I had to find that elusive creature, the good practice, whatever that was. I had begun to have nightmares, tossing and turning as I struggled in vain to fit a six unit, three pontic, fixed porcelain bonded bridge, or to extract a distally impacted wisdom tooth from a bricklayer who was built like the proverbial outhouse. More often I couldn't shift an upper central incisor with gross periodontal disease from a sweet little 70-year-old lady. It was always with relief that I would wake up, albeit in a cold sweat, and realise that it had all been a dream. However, I was scared that one day it could become a terrifying reality.

In the dark

Horror stories had filtered back to me along the lines of: 'Well, of course, he started working for this guy 6 months ago and he's still only grossing a quarter of target. What's more, he's on 35% with a one-year, 25-mile binding out agreement'. What did it mean? I could only guess from the suitably grave faces around me.

More mundanely, I was used to having 50 contemporaries around me in the busy hospital environment; the thought of being a lone practitioner with no easy contact with others was at best depressing.

I was concerned that I wouldn't cope with the speed necessary to treat patients ethically and to earn a living. I didn't believe it was possible to cut a crown in under 2 hours, and I had visions of finishing work for lunch one-and-a-half hours after I was supposed to have started for the afternoon.

Finally and most worryingly, particularly as finals approached with the attendant sudden interest in hitherto unexplored text books, I became aware that there were huge gaps in my knowledge. I came to the realisation that 5 years of study and training had merely equipped me with the basics. What's more, dentistry changes, and of course one cannot practise dentistry as one learned it for very long before one's methods become out of date.

I was becoming more and more aware of my inadequacies. How was I going to cope with being an independent practitioner?

It finally dawned on me that many of my problems could be eased or even solved by joining a VTS practice. I decided to take my opportunity and accept the job. The training practice which I eventually joined came under the auspices of the scheme in Trent.

Four days a week were spent treating patients at the

practice, and one in attending a day release course at Queens Medical Centre in Nottingham. To give some idea of the scope of the work covered and its variety, I will briefly list some of what we covered.

The day release course itself took the form of seminars in which discussion and participation were encouraged. The subjects covered ranged from elaboration on undergraduate topics—such as cross-infection control, referrals, consultation techniques, and anaesthesia, analgesia and sedation—through to many topics only touched on before. These included the work of the district dental officer and the dental reference officer, how accountants and solicitors can help in the running of your practice, the science (or some would say art) of equipment maintenance, health and safety at work, and the more peripheral aspects of a GDP's work such as hypnosis and clinical photography.

Several visits were arranged, some of them overnight. These included some to the local Family Practitioner Committee, the Dental Practice Board, the headquarters of the BDA and the GDC, and one to a large dental laboratory. We also spent an interesting and mutually rewarding day in the surgeries of local general medical practitioners, accompanying them on their rounds.

Regular problem-solving sessions completed the timetable of work. These consisted of brief presentations by each trainee of the notes, radiographs and study models of a patient of their choice. This was followed by a discussion, which was sometimes heated, on the best way to treat him. A longer and more detailed case report on a patient for whom we had completed treatment was presented at the final session of the year. We also kept a record of our improving abilities in conjunction with our trainers, which enabled us to monitor our progress.

I wanted to be able to speak with more than my own opinion to go on when I discussed to what extent the scheme I have outlined above succeeded in overcoming the obstacles that the new graduate faces. To this end, I sent out a questionnaire at the end of the vocational training course to the other trainees on the Trent scheme. My aims were to discover their reasons for participating and their degree of satisfaction with their decision to do so. Of the 12 that I contacted, nine replied. As the sample is small, I intend only to draw trends from it, as statistical analysis would be misleading.

Questions and answers

My first question was aimed at finding out their reasons for joining the course in the first place. Answers included an awareness of the need for postgraduate education; job stability was mentioned, as was the help that was available in the surgery if necessary. For one person the job they liked just happened to be VTS, and for another the 4-day clinical week was a factor—if not a very worthy one!

However, four reasons were mentioned most often; the ease transition from the nannying of the hospital training system to the challenge of independent practice, the security of knowing one was in an ethical practice which had been vetted by an experienced dentist, the guaranteed income, and the opportunity to meet other recently graduated dentists were all popular reasons for joining the scheme. The trainees expectations of the day release course centred on the belief that they would further extend their dental education, and fulfil one of the previously mentioned needs—that of meeting their peers. It is heartening to see that seven of the nine respondents believed that the course fulfilled their expectations at least quite well.

I asked for any criticism of the course, and did indeed receive several suggestions on what could have been improved. However, of the 12 separate comments made, only five were repeated, and only one of these was repeated more than once. This was that more could have been put into the day release course, at the expense of some of the material that was a repeat of undergraduate work or had been mentioned in connection with another seminar session. When asked what topics in particular could have been added, the hands-on type of course, together with more on oral surgery were mentioned several times. Consistent praise was given to the opportunity to meet other new graduates, and to the interesting nature of many of the seminars.

I then asked how helpful each individual facet of the whole scheme had been.

The weekly meetings at Nottingham, the visit to the DPB, and the visit to a general medical practitioners surgery were all quoted by at least five of the respondents as being very useful. The trainer-trainee discussions, the overnight courses, and the opportunity to assess oneself on video in the surgery during consultations were each found at least fairly useful by five or more respondents.

The penultimate question was on the various forms of assessment used. Respondents were asked to give each one a score out of five. The runaway winner in this was the presentation of the case reports which took place at the end of the year. The progress record filled in with the trainers also received a creditable response. Less well received were the research projects and the log diaries we kept of the treatments which we had performed.

Perhaps the most telling response was in answer to the final question: 'Overall, are you glad that you participated on the course?'. Nine out of nine said yes.

On examining the results of the survey, there was a great variety noticeable in people's expectations and in their view of the reality of the course. However, certain factors did stand out.

Shared concerns

It seemed that my own concerns of 18 months ago were indeed mirrored to a certain extent by the other trainees. The oft-expressed wish for more advanced education together with a support system of experienced dentists to fall back on, and like-minded peers to have a grouse with, underline the insecurity felt by many as they faced the traumatic transition to general practice.

Something for the powers that be to consider when planning mandatory vocational training is the high praise for the salaried system. This was quoted by many as a reason for joining the course in the first place and common sense tells us that it must be a major factor in removing the pressure on a less than confident and inexperienced practitioner.

High praise was also reserved for the aforementioned problem-solving sessions, much commended in the morning coffee break post mortems. These could certainly be extended —the problem-solving sessions, that is, not the coffee breaks!

The knowledge that one was in an ethical, well-run practice was a very important factor for many. If vocational training is to become mandatory, it is important that the high standards so commendably set and achieved in 1988 must be upheld if graduates' faith in the system is to be maintained.

There was little consensus on any poor aspects of the scheme. Indeed, in some cases what one person listed under criticisms another would place under praises. Working under the assumption that you can't please all of the people all of the time, we can conclude that the VTS scheme was probably satisfactory in meeting a lot of its trainees' requirements.

To what extend do I feel my time on the vocational training scheme helped me in my transfer to independent practice?

As an associate in the practice in which I did my vocational training I now look back and see how very fortunate I was to have landed a VTS job last year. Over the course of my time in the surgery I lost count of how many times, particularly in the early months, I called on the help that I knew was readily available.

Useful tips

There was no need to pretend that I knew what to do when faced with a tricky situation. There was no need to struggle on with a particular item of treatment which wasn't going well. I know I would have been embarrassed by asking for help had it not been expected that I would. Sometimes all that was needed was for me to gain the expertise I needed from experience, and eventually I called on my trainer less and less. But many times a useful tip or comment made the job a hundred times easier instantly. At dental school, with the welfare of several students and many patients at stake, the more minor faults in technique may go undetected by the busy senior house officer or registrar in charge. The one-to-one basis of the trainer-trainee relationship, however, means that these minor faults are quickly spotted and rectified.

The only time I remember my trainer not being able to work a miracle and bail me out was on the occasion when my try-in dentures wouldn't fit. We eventually worked out that this was due to a mix up in the laboratory and the dentures that I was attempting to fit were actually for the patient after next.

One of my most useful half-hours was spent chatting over coffee with a friend who did a lot of endodontics with her trainer. She passed on several tips which, in no great expectations of success, I tried. To my amazement, both the quality and the speed of my endo improved dramatically. Far fewer patients returned with symptoms after initial emergency treatment for pain. Later, my own trainer taught me his method for lateral condensation of GP, which resulted in far better obtunded root canals than I had been used to obtaining. I can confidently say that as a result of these two short sessions, my root canal treatments improved to such an extent that they alone made my attendance on the VTS course worthwhile. I would not have learned this working on an associate basis, tucked away in my own surgery.

Financially, the VTS trainee is undoubtedly better off in the initial months than his contemporary GDP, even if the balance does swing back over the later months. The consequent lack of pressure on time was in no way better appreciated than when I was working on complex conservation or oral surgery. These were things I had done relatively little of at dental hospital and I was worried enough about the technicalities of them without having to worry about timing as well.

Growing confidence

Naturally, as time went on I grew more accomplished and am now quite happy to cut crown preps under pressure. But on first qualifying, it was an altogether different story. I'm sure much blood, sweat and tears (and not all of it mine) were saved in those early months.

In conclusion, I feel I must agree with the prediction that the VTS will become mandatory in years to come. The leap from the structured environment of the dental school to the cut and thrust of the everyday work in general practice is a chasm so wide, especially to the young graduate who has to cross it, that it is surprising that it is only comparatively recently that anyone has thought of building a bridge.

Ongoing assessment of the trainees' thoughts and criticisms year by year will be essential if the scheme is to develop from the promising embryo it now is to become a fully-fledged postgraduate training scheme.

Constructive criticism and praise must be taken on board, and if, as with clinical dentistry, the course moves with the times and keeps up to date, I am sure that the young dentists of years to come will greatly appreciate the opportunity to assimilate the knowledge that they need to become competent general dental practitioners. They may even be astounded that there was ever a time when it was not considered essential.

H. G. Wells once said that human history becomes more and more a race between education and catastrophe. The vocational training scheme is certainly doing its bit to ensure that education wins.

34

My Year in Vocational Training: Highlights and Horrors

C. Chavasse

I am now drawing towards the end of my year in Eastbourne as a vocational trainee on the Brighton scheme. Eastbourne is a busy seaside resort on the south coast. Rather a lot of 'grannies' reside there but contrary to popular belief I do not spend all day, every day making and repairing dentures! The work is varied and interesting and I am spending a happy year consolidating my knowledge and becoming more confident at practising dentistry. Here are a few of the best and worst parts of the year.

✗ Chelwood Gate, first visit: this was a residential weekend course that took place in a conference centre in the middle of nowhere (but not far from East Grinstead). All the 'fresher' trainees from several schemes gathered together there to get to know each other. The idea was a good one but the lectures weren't particularly inspiring, apart from the ambulance man who came and retaught us all cardio-pulmonary resuscitation (CPR). There was also an acutely embarrassing episode on the first evening of group therapy—confiding in a circle of complete strangers on one's thoughts and personal details. Perhaps a party would have been a better way of meeting the others.

✓ Chelwood Gate, second visit: this was much better—everyone knew everybody and we had some excellent speakers. Mr Frank Taylor stirred up a great deal of excitement when talking about practice management, but unfortunately was stopped in his tracks by limited time almost before he had started; Mr Simon Wilsher kept us magnificently entertained with an evening of modified pelmanism, or 'how to improve your memory'!

All I can remember is visions of my DSA on horseback and enormous probes being hurled at me by swarming giant black and yellow bees. If you don't understand, go to the lecture!

✓ Some Fridays (our day release) we would all go for a pub lunch in between lectures when one could catch up on news with old dental buddies, compare notes on one's practice facilities, systems, bosses and so on, and generally relax and think TGIF (thank God it's Friday!).

✗ Bouncing into the practice on a Monday morning full of enthusiasm and good intentions to implement the fantastic ideas on various techniques, materials, or safety aspects (that one had learnt the previous Friday) to be brought down to earth with a large bump because the idea was too expensive, too time-consuming, too longsighted or too much of a change when 'what was wrong with the old way!'

✓ We had some excellent clinical lectures and study days, of which the following are just a few:
● Endodontics by Mr C. Nehammer.
● Rubber Dam (Dam it, it's easy!) by Mr K. Marshall. These two subjects were complete grey areas to me even when I qualified and therefore not exactly my favourite aspects of dentistry; the above-mentioned study days completely reformed that prejudice, as the light dawned and they were both made to seem easy. I can and do use both procedures

quickly and efficiently (well, one hour instead of 3 for RCT!)
● Four-handed dentistry by Mr Ellis Paul.
● Crowns and veneers by Mr R. Goulden—the latter of which I had just missed being taught at dental school.

We also had excellent lectures on finance, computing, marketing and management, occlusion, and interesting visits to local practices and to Brighton Hospital.

✗ Endless lectures on the FPC, GDS, DRS, LDC, DHA, CHC, GDS . . . and so on, which, although it is vital to know about and to understand the role of each, why is no one able to make them sound interesting?

✓ Being salaried. This enabled me to gain speed gradually without having to worry about the wage packet at the end of the month. This was particularly good as I was building up a new book. The other excellent advantage was that I was able to buy a property which would have been virtually impossible as a newly-qualified person with no financial record and an irregular income as on the fee-per-item system.

✗ The thought of hitting reality next year. In general practice I will have to gross nearly twice as much as I do now to earn the same amount.

Those were some of the good and not so good points about the scheme in general. Here are a few in and around the practice in Eastbourne.

✓ The practice is literally 10 yards from the sea.
✗ Eastbourne is out in the sticks, the road access to it is appalling.

✓ The staff were all very pleasant and easy to get on with.
✗ They were continually changing. Two nurses, three receptionists, three cleaners and a technician left while I was there. OK, the circumstances were exceptional and some of the changes were for the good, but there never seemed to be a 'normal' day when all were present and all the equipment worked properly. It was certainly an insight into the nightmare of running a largish practice.

✓ I had such a domineering 19-year-old nurse when I first started in the practice that *I cleared* up after *her*!
✗ I then trained my own nurse—this was good to a certain degree because she knew less about practice than I did, so she did as she was asked, but it is not as easy as it might appear to train a nurse, and I wasn't really experienced

enough to do so, therefore she is probably only half-trained.

✓ For my first 8 months I went round to my principal's house for supper and my weekly tutorial. I was able to sort out many problems and worries by doing this. It also meant an evening out when I knew no one else in Eastbourne, which was really very much appreciated.

✗ I found it difficult to communicate with the boss about anything other than immediate clinical problems when *within* the practice walls.

✓ Being given use of the computer and some clinical devices to experiment with.

✗ Having a 50-year-old chair that dropped 2 mm with a dull thud every 15 seconds so that a patient, having begun treatment at the correct Ellis Paul height, would be down somewhere near my ankles by the time I was burnishing the amalgam restorations!

I also had no radiograph machine in my room which caused chaos most days—the patients had to either play musical chairs or wait 20 minutes for their bitewings!

✓ The first surgical I completed *on my own*!

✗ The preceding six!—from which I had to be rescued and then ran an hour behind for the rest of the morning.

Altogether I have had a very exciting, enlightening year which I have enjoyed very much.

I think one of the most important messages that was continually put over to us was to carry on learning, to carry on going to lectures and study days even when the year was up, in order to keep up to date and to keep one's brain functioning and questioning, and to keep life interesting so that dentistry can be enjoyable for life.

I highly recommend vocational training to all newly qualified dentists, even if you have done house jobs; it is excellent.

Appendices

Appendix 1   Suggested reading list

Basic sciences

Berkovitz B K, Holland G R, Moxon B J. *A colour atlas and textbook of oral anatomy*. London: Wolfe Medical, 1978.
Adams D. *Essentials of oral biology*. Edinburgh: Churchill Livingstone, 1981.
Green J. *An introduction to human physiology*. 4th ed. Oxford: Oxford University Press, 1976.
Reese A. *The principles of pathology*. Bristol: Wright, 1981.

Oral diseases

Cawson R. *Essentials of dental surgery and pathology*. 4th ed. Edinburgh: Churchill Livingstone, 1984.
Scully C. *Patient care. A dental surgeon's guide*. 2nd ed. London: British Dental Journal, 1990.

Periodontology

Strahan J D, Waite I M. *A colour atlas of periodontology*. London: Wolfe Medical, 1978.

Jurisprudence

British Dental Association. *Ethical and legal obligations of practitioners*. London: BDA.
General Dental Council. *Professional conduct and fitness to practise*. London: GDC.
Seear J, Walters L. *Law and ethics in dentistry*. 3rd ed. Bristol: Wright, 1990. Dental Practitioner Handbook 19.

Conservation

Gainsford I D. *Silver amalgam in clinical practice*. 2nd ed. Bristol: Wright, 1976. Dental Practitioner Handbook 10.
Smith B G N. *Planning and making crowns and bridges*. London: Martin Dunitz, 1986.
Stock C J R, Nehammer C. *Endodontics in practice*. 2nd ed. London: British Dental Journal, 1990.
Rowe A H R, Alexander A G, Johns R B. *Clinical dentistry*. London: Class Publishing, 1989. Companion to Dental Studies 3.

Prosthetics

Basker R M, Davenport J C, Tomlin H R. *Prosthetic treatment of the edentulous patient*. London: MacMillan, 1983.
Neill D, Walter J D. *Partial denture prosthesis*. 2nd ed. Oxford: Blackwell, 1983.
Hickey J C, Zarb G A, Bolender C L. *Boucher's prosthodontic treatment for edentulous patients*. 9th ed. St Louis: C V Mosby, 1985.

General practice

Bennett M. *General dental practice*. Edinburgh: Churchill Livingstone, 1978.
Tay W M, Rear S. *General dental treatment*. Edinburgh: Churchill Livingstone, 1982.

Pharmacology

Cawson R, Spector R G. *Clinical pharmacology in dentistry*. Edinburgh: Churchill Livingstone, 1989.
British national formulary. London: British Medical Association and The Pharmaceutical Society of Great Britain.

Radiology

Beeching B. *Interpreting dental radiographs*. Guildford: Update Publications, 1980.
DH Standing Dental Advisory Committee. Radiation protection in dental practice. London: Department of Health.

Oral surgery

Seward G R, Harris M, McGowan D A. *An outline of oral surgery. Parts 1 and 2*. Bristol: Wright, 1975. Dental Practitioner Handbook 11.

Orthodontics

Mills J R E. *The principles and practice of orthodontics*. Edinburgh: Churchill Livingstone, 1988.

Paedodontics

Wright G Z, Starkey P E, Gardner D E. *Child management in dentistry*. Bristol: Wright, 1987. Dental Practitioner Handbook 28.
Andlaw R J, Rock W P. *A manual of paedodontics*. Edinburgh: Churchill Livingstone, 1987.

Occlusion

Wise M D. *Occlusion and restorative dentistry for the dental practitioner*. 2nd ed. London: British Dental Journal, 1986.

Dental photography

Wander P, Gordon P. *Dental photography*. London: British Dental Journal, 1987.

Dental materials

Combe E C. *Notes on dental materials*. Edinburgh: Churchill Livingstone, 1986.
Wilson H J, McLean J W, Brown D. *Dental materials and their clinical application*. London: British Dental Journal, 1988.

Medicine

Scully C, Cawson R A. *Medical problems in dentistry*. 2nd edition. Bristol: Wright, 1987.

Prevention—dental health education

Murray J J. *The prevention of dental disease*. Oxford: Oxford University Press, 1989.
Elderton R J. *Positive dental prevention. The prevention in childhood of dental disease in adult life*. London: Heinemann, 1987.
Murray J J, Rugg-Gunn A J. *Fluorides in caries prevention*. Bristol: Wright, 1976. Dental Practitioner Handbook 20.
Jacob M C, Plamping D. *People, patients and prevention. The practice of primary dental care*. Bristol: Wright, 1989.
Locker D. *An introduction to behavioural science and dentistry*. London: Routledge, 1989.
Sheiham A, Cowell C R. *Promoting dental health*. London: King's Fund, 1982.
Ewles L, Simnett I. *Promoting health. A practical guide to health education*. West Sussex: John Wiley & Sons, 1985.

Handicapped

Hunter B. *Dental care for handicapped patients*. Bristol: Wright, 1987. Dental Practitioner Handbook 36.

Gerodontics

Bates J F, Adams D, Stafford G D. *Dental treatment of the elderly*. Bristol: Wright, 1984. Dental Practitioner Handbook 35.
Holm-Pederson P, Loe H. *Geriatric dentistry: a textbook of oral gerontology*. Copenhagen: Munksgaard, 1986.
Tryon A F. *Oral health and ageing: an interdisciplinary approach to geriatric dentistry*. Littleton, Mass: PSG Publishing, 1986.

Community dental health

Slack G. *Dental public health: an introduction to community dental health*. 2nd ed. Bristol: Wright, 1981.

Appendix 2 Addresses of regional postgraduate dental deans and advisers

ENGLAND

East Anglia region
Regional Director of Postgraduate Dental Education,
Ipswich Hospital,
Heath Road Wing,
Heath Road,
Ipswich IP4 5PD.

Mersey region
Postgraduate Dental Dean,
University of Liverpool,
School of Dentistry,
Pembroke Place,
P.O. Box 147,
Liverpool L69 3BX.

Northern region
Postgraduate Dental Dean,
The Dental School,
The University of Newcastle upon Tyne,
Framlington Place,
Newcastle upon Tyne NE2 4BW.

North Western region
Dean of Postgraduate Dentistry,
University Dental Hospital,
Higher Cambridge Street,
Manchester M15 6FH.

Oxford region
Regional Director of Postgraduate Dental Education,
Department of Oral Surgery,
John Radcliffe Hospital,
Headington,
Oxford OX3 9DU.

South Western region
Dental Postgraduate Dean,
University of Bristol Dental School,
Canynge Hall,
Whiteladies Road,
Bristol BS8 2PR.

Thames region
Regional Postgraduate Dental Dean,
British Postgraduate Medical Federation,
33 Millman Street,
London WC1N 3EJ.

Trent region
Regional Postgraduate Dental Dean,
Transitional Training Unit,
University of Sheffield,
School of Clinical Dentistry,
Charles Clifford Dental Hospital,
Wellesley Road,
Sheffield S10 2SZ.

Wessex region
Regional Director of Postgraduate Dental Studies,
Wessex Regional Health Authority,
Highcroft,
Romsey Road,
Winchester,
Hampshire SO22 5DH.

West Midlands region
Postgraduate Adviser in Dentistry,
The Dental School,
The University of Birmingham,
St. Chad's Queensway,
Birmingham B4 6NN.

Yorkshire region
Postgraduate Dental Dean,
University of Leeds,
School of Dentistry,
Clarendon Way,
Leeds LS2 9LU.

SCOTLAND
Postgraduate Adviser in Dentistry,
Dental Department,
Aberdeen Royal Infirmary,
Aberdeen AB9 2ZB.

Postgraduate Adviser in Dentistry,
Dental School,
Park Place,
Dundee DD1 4HN.

Mr Hew Mathewson,
Postgraduate Dental Adviser for South East Scotland,
Pfizer Foundation,
Hill Square,
Edinburgh EH8 9DR.

Postgraduate Adviser in Dentistry,
West of Scotland Centre for Postgraduate Dental Education,
378 Sauchiehall Street,
Glasgow G2 3JZ.

WALES
Subdean of Postgraduate Studies (Dental),
University of Wales College of Medicine,
Dental School,
Heath Park,
Cardiff CF4 4XY.

NORTHERN IRELAND
Postgraduate Adviser in Dentistry,
School of Clinical Dentistry,
Royal Victoria Hospital,
Grosvenor Road,
Belfast BT12 6BP.

Index